Advanced Level Biology

A2 Biology with Stafford

Unit Five: Physiology of Exercise & Nervous Coordination

Paper reference: 6BIO5

Typeset & layouts by   : Mohamed Sobir
Cover designed by   : Mohamed Sobir
Printed in Maldives by   : Copier Repair
Published by   : Author publisher
Date   : January 2011

ISBN: 978-81-910705-4-5

Contact details of the author:

Stafford Valentine Redden
Near Hindhustan Oil Mill,
Ramchandrapur, Jatni P.O,
Orissa, India, 752050
staffordv@yahoo.com
+960 7765507

Dedicated to my all my loving students who
have been a continuous source of inspiration

Contents

Chapter One
Muscle Structure & Function   1

Chapter Two
Cellular Respiration   10

Chapter Three
Electrocardiograms (ECGs)   18

Chapter Four
Regulation of Heartbeat Rate & Breathing Rate   21

Chapter Five
Spirometry   24

Chapter Six
Feedback Mechanisms   28

Chapter Seven
Gene Induction   30

Chapter Eight
Benefits & Risks of Exercise   32

Chapter Nine
Laproscopy, Sport Injuries & Ethics of Using Drugs   34

Chapter Ten
Photoreception in Plants   39

Chapter Eleven
Structure & Function of Neurones   41

Chapter Twelve
Synaptic Transmission   47

Chapter Thirteen
Photoreception in Mammals   51

Chapter Fourteen
Comparison of Coordination in Plants & in Animals   55

Chapter Fifteen
The Human Brain   56

Chapter Sixteen
Investigating Brain Structure & Function   58

Chapter Seventeen
Critical Window in Development of Vision   60

Chapter Eighteen
Nature & Nurture in Brain Development   65

Chapter Nineteen
Habituation   69

Chapter Twenty
Ethics of Using Animals in Medical Research   71

Chapter Twenty One
Parkinson's Diseases & Depression   72

Chapter Twenty Two
The Human Genome Project & Development of New Drugs   75

Topic Seven
Run for Your Life

Topic Eight
Grey Matter

CHAPTER ONE
MUSCLE STRUCTURE AND FUNCTION

Learning outcomes: by the end of this chapter you should be able to
Edexcel Syllabus Spec 1: *Demonstrate knowledge and understanding of the practical and investigative skills identified in numbers 4 and 5 in the table of How Science Works on page 13 of this specification.*
Edexcel Syllabus Spec 2: *Describe the structure of a muscle fibre and explain the structural and physiological differences between fast and slow twitch muscle fibres.*
Edexcel Syllabus Spec 3: *Explain the contraction of skeletal muscle in terms of the sliding filament theory, including the role of actin, myosin, troponin, tropomyosin, calcium ions (Ca2+), ATP and ATPase.*
Edexcel Syllabus Spec 4: *Recall the way in which muscles, tendons, the skeleton and ligaments interact to enable movement, including antagonistic muscle pairs, extensors and flexors.*

Structure of a sarcomere – The functional unit of muscles

The functional unit of contraction in a muscle is the **sarcomere**. Muscle cells contain many sarcomeres arranged in parallel. The muscle cell takes on a characteristic banded appearance because of the regular arrangement of the sarcomeres. This is called **striation**.

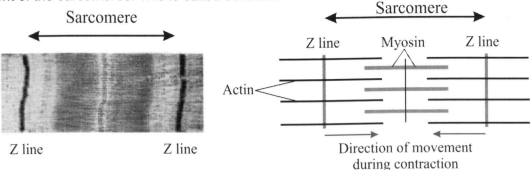

Fig. 1.1 A sarcomere: Note the striated appearance of the muscle

The sarcomere contains overlapping actin and myosin. The myosin is often called the **thick filament** because the myosin heads make it appear thick. The actin is, therefore, the **thin filament.**

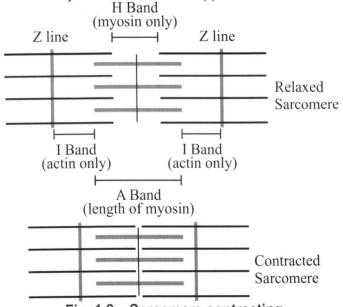

Fig. 1.2 – Sarcomere contracting

Z line: a disc to which actin filaments are attached. The sarcomere extends from one Z disc to the next.
H Band: Region of sarcomere with myosin only; no actin present.
I Band: Region of sarcomere with actin only; no myosin present.
A Band: Represents the length of myosin filaments. May or may not overlap with actin.

When a sarcomere contracts the **H band and I band shrink**, but the **A band remains unchanged**. The actin **slides** over the myosin towards the H zone. This **pulls the Z discs closer to each other and the sarcomere contracts**, as shown in fig. 1.2.

The sarcomeres are arranged in a chain to form a myofibril. The myofibrils are bundled together and surrounded by a sarcolemma (muscle cell membrane) to form a muscle fibre or muscle cell. The muscle fibre contains sarcoplasm (muscle cytoplasm), nuclei, sarcoplasmic reticulum and mitochondria.

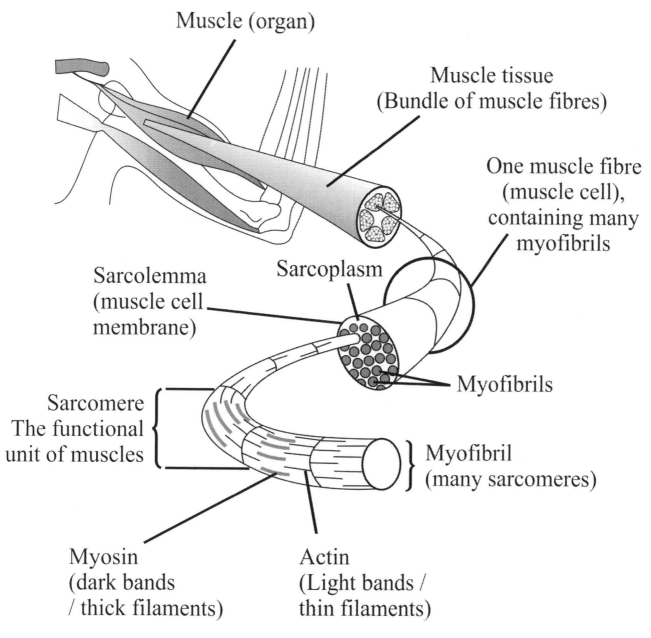

Muscle (organ)

Muscle tissue
(Bundle of muscle fibres)

One muscle fibre
(muscle cell),
containing many
myofibrils

Sarcolemma
(muscle cell
membrane)

Sarcoplasm

Myofibrils

Sarcomere
The functional
unit of muscles

Myofibril
(many sarcomeres)

Myosin
(dark bands
/ thick filaments)

Actin
(Light bands /
thin filaments)

Fig. 1.3 – Structure of muscle

The muscle fibres (cells) are further arranged into larger bundles to form muscle tissue. The combination of muscle tissue along with nerve cells, blood and connective tissue forms an organ (muscle).

Levels of organisation in muscles

Muscle (organ - Muscle tissue, connective tissue, blood vessels and neurones)

⬇

Muscle tissue - many muscle fibres / cells held by connective tissue

⬇

Muscle fibre (cell) - contains many myofibrils in sarcoplasm.

⬇

Myofibril - made up of many sarcomeres

⬇

Sarcomere - made up of actin and myosin myofilaments

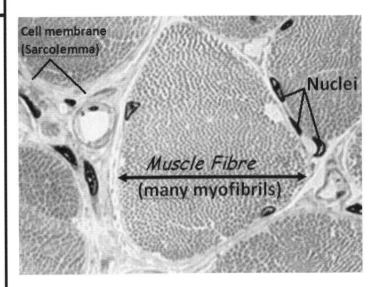

Fig. 1.4 – Arrangement of myofibrils

In order to understand the physiology of muscle contraction, it is necessary to understand the structure and role of actin, myosin, troponin, tropomyosin, calcium ions, ATP and ATPase. The diagrams below provide some useful information.

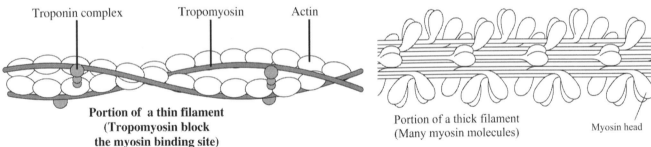

Portion of a thin filament
(Tropomyosin block
the myosin binding site)
Fig.1.5 – Actin (Thin filament)

Portion of a thick filament
(Many myosin molecules)   Myosin head
Fig.1.6 – Myosin (Thick filament)

Actin (thin filaments): It contains myosin binding sites. It attaches to the myosin heads and slides towards the H band. The sliding of actin pulls the Z disc and causes the sarcomere to contract.

Tropomyosin: blocks the myosin binding sites on actin filaments when the muscle is relaxed. This prevents myosin from binding to actin.

Troponin: combines with calcium ions and causes tropomyosin to shift. This exposes the myosin binding sites on actin.

Myosin (thick filaments): binds to actin and pulls it towards the H zone.

Calcium ions: Released from the sarcoplasmic reticulum (modified smooth endoplasmic reticulum) in response to a nerve impulse. The calcium ions bind to troponin and expose the myosin binding sites on actin filaments.

The fig.1.7, 1.8 and 1.9 are models used to illustrate the role of troponin, tropomyosin and calcium ions in muscle contraction. Note that the actual arrangement of troponin and tropomyosin is shown in fig. 1.5.

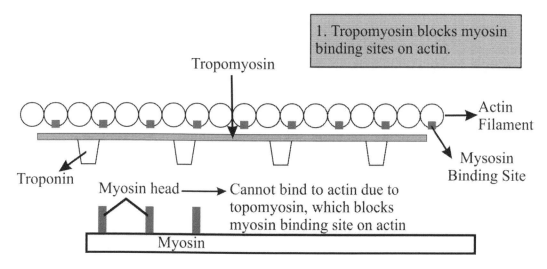

Fig. 1.7 – Tropomyosin blocks myosin binding sites on actin

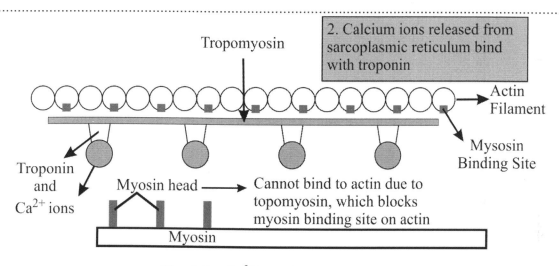

Fig. 1.8 – Ca^{2+} ions bind to troponin

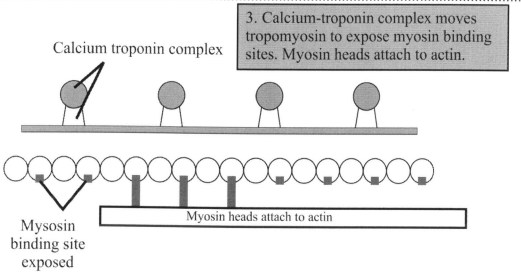

Fig. 1.9 – myosin binding sites on actin are exposed and myosin binds to actin

The process by which the thin actin filaments are pulled in towards each other by the myosin is called the sliding filament theory. The sliding of actin over myosin makes muscles contract. The sequence of events, which bring about muscle contraction is illustrated in fig. 1.10.

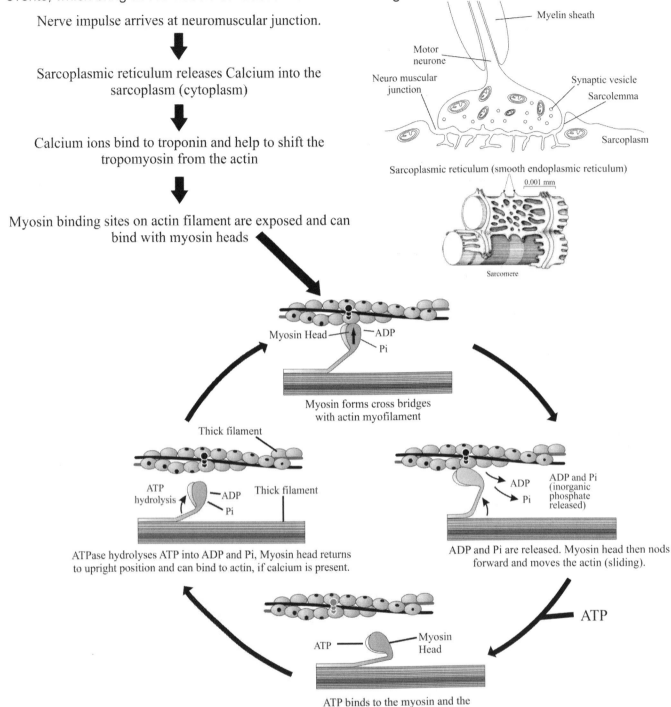

Nerve impulse arrives at neuromuscular junction.

⬇

Sarcoplasmic reticulum releases Calcium into the sarcoplasm (cytoplasm)

⬇

Calcium ions bind to troponin and help to shift the tropomyosin from the actin

⬇

Myosin binding sites on actin filament are exposed and can bind with myosin heads

Myosin Head —ADP
Pi

Myosin forms cross bridges with actin myofilament

Thick filament

ATP hydrolysis —ADP
Pi
Thick filament

ATPase hydrolyses ATP into ADP and Pi, Myosin head returns to upright position and can bind to actin, if calcium is present.

ADP
Pi
ADP and Pi (inorganic phosphate released)

ADP and Pi are released. Myosin head then nods forward and moves the actin (sliding).

ATP

ATP —Myosin Head

ATP binds to the myosin and the head detaches from actin.

Fig. 1.10 – Sliding filament theory of muscle contraction

Role of ATP and ATPase

ATP is required to detach myosin from actin. Hydrolysis of ATP (into ADP and P_i) **energises the myosin head**, so that it can bind to actin again and undergo another contraction cycle.

If ATP levels drop (assuming Ca^{2+} is present) the myosin stays attached to the actin and the muscle stays permanently contracted. This is what causes **rigor mortis.**

ATPase is an enzyme which hydrolyses ATP into ADP and P_i. The ADP and P_i remain attached to myosin and cause it to return to its upright position. The myosin can then bind with actin, if calcium is present.

The chart below summarises the events that occur during muscle contraction

1. Nerve impulse arrives at a neuromuscular junction and releases neurotransmitters.

⇩

2. Calcium ions (Ca^{2+}) are released from the sarcoplasmic reticulum into the sarcoplasm.

⇩

3. Calcium ions (Ca^{2+}) bind to troponin.

⇩

4. Troponin molecules and the attached tropomyosin shift their position, exposing myosin binding sites on the actin filaments.

⇩

5. Myosin heads (containing ADP and Pi) bind with myosin binding sites on the actin filament, forming cross-bridges.

⇩

6. When the myosin head binds to the actin, ADP and Pi on the myosin head are released.

⇩

7. The myosin changes shape and nods forward, causing the actin to **slide** over the myosin. The actin slides towards the H zone.

⇩

8. An ATP molecule binds to the myosin head, causing it to detach from the actin. Lack of ATP prevents detachment, leading to rigor mortis or stiffening of muscles.

⇩

9. An ATPase on the myosin head hydrolyses the ATP, forming ADP and Pi.

⇩

10. The myosin head changes shape and returns to its upright resting position.

⇩

11. The cycle can be repeated again if Ca^{2+} ions are present.

When a muscle relaxes it is no longer being stimulated by nerve impulses. Calcium ions are actively pumped out of the muscle sarcoplasm, using ATP. The troponin and tropomyosin move back, once again blocking the myosin binding site on the actin.

Slow twitch and fast twitch fibres

Muscle fibres that are adapted to undergo very rapid and powerful contractions are called as fast twitch fibres, where as fibres which are adapted to undergo slow and weak contractions are called as slow tiwtch fibres.

The muscle type of a cheetah or a gazelle will be predominantly **fast twitch**, whereas the muscle of a camel or an elephant will be predominantly **slow twitch**. Muscle type in humans is predominantly one or the other due to inherited alleles. However, different training programmes can cause the proportion of either type to change slightly. Skeletal muscles are a mixture of both types of fibre, but the proportions depend on what the muscle is used for.

The structural and physiological differences between the two types of fibres are discussed in the table that follows.

Feature	Slow twitch fibres or Type I (Slow Oxidative)	Fast twitch fibres or Type II (Glycolytic fibres)
Diameter	Have a narrow diameter, as they have few myofibrils per fibre.	Thick fibres. Many myofibrils to generate maximum force.
Force generated	Low force, Weak contractions.	High Force, Strong contractions.
Speed of contraction	Contracts slowly.	Contracts rapidly.

Type of respiration	Slow twitch fibres are adapted to **respire aerobically**. This ensures that there is a continuous steady supply of ATP and lactic acid formation is minimal.	Fast twitch fibres **need a rapid supply of ATP**. The **aerobic pathway is too long** and involves too many reactions to produce ATP. On the other hand, **anaerobic respiration produces ATP during glycolysis only**, which is a much **shorter pathway** and can **supply ATP rapidly**.
Resistance to fatigue	Very resistant to fatigue (tiredness), as they **do not produce lactic acid** easily.	Low resistance to fatigue. They tire or cramp up easily due to **lactic acid formation**. The lactic acid reduces the pH within the muscle cells and inhibits respiratory enzymes. This results in complete lack of ATP leading to muscle cramps.
Myoglobin content	**Myoglobin** is a **dark red** respiratory pigment which stores oxygen in the muscles. It is made up of a single polypeptide chain associated with one Haem (iron containing) group. It has a very high affinity for oxygen. It can store oxygen in the form of oxy-myoglobin. When oxygen concentration in muscles starts to decrease, myoglobin releases the oxygen for the muscle to continue aerobic respiration. Slow twitch fibres are referred to as **red muscle** as it contains lots of myoglobin.	Fast twitch fibres are referred to as white muscle and is light pink in colour. It has very little myoglobin and is adapted for anaerobic respiration. If myoglobin was present in abundance, then the muscle will continue aerobic respiration and the rate of ATP production would be **too slow to support rapid contraction**. **Note:** the anaerobic respiration pathway is <u>very short</u> when compared to the aerobic pathway. This enables **ATP to be produced very rapidly** for muscle contraction.
Mitochondria	Have many mitochondria, which facilitate aerobic respiration.	Have few mitochondria as muscles undergo anaerobic respiration, which occurs in the cytoplasm.
Sarcoplasmic reticulum	Few sarcoplasmic reticulum, which store and release Ca^{2+} ions for muscle contraction.	Have lots of sarcoplasmic reticulum, which release plenty of Ca^{2+} ions for frequent muscle contraction.
Glycogen concentration	Have low glycogen content, as glucose is supplied continuously by numerous capillaries.	Have Lots of glycogen, which is converted into glucose and used instantly for respiration. **Note:** Conversion of glycogen to glucose, in the liver, and subsequent transfer to the muscles would be too slow to meet the rapid demand for glucose during anaerobic respiration.
Blood capillaries	Have numerous capillaries, which provide a continuous supply of oxygen and glucose for aerobic respiration.	Have few capillaries, which result in reduced oxygen supply and anaerobic respiration. The pathway for anaerobic respiration is very short (Glycolysis only) and generates ATP rapidly.
Creatine phosphate	Creatine phosphate is found in low concentration as these fibres undergo aerobic respiration.	Have a high concentration of creatine phosphate which reacts with ADP in a single step to form ATP rapidly for muscle contraction. Creatine Phosphate + ADP $\longrightarrow$ ATP + Creatine
ATPase activity	ATPase activity is slow. ATPase hydrolyses ATP and helps the myosin head to re-cock rapidly.	ATPase activity is fast. Helps myosin to re-cock rapidly and bring about another round of contraction rapidly.

Joints and movement

Joints are the areas where two or more bones meet. Most joints are mobile, allowing the bones to move.

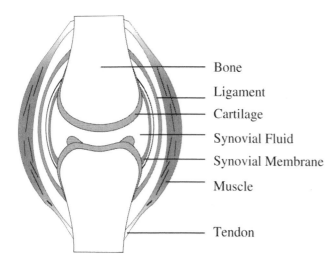

Fig. 1.11 – structure of a synovial joint

Joints consist of the following:
- **cartilage** - at the joint, the bones are covered with cartilage (a connective tissue), which is made up of cells and fibres and is wear-resistant. Cartilage helps reduce the friction of movement and absorbs mechanical shock.
- **synovial membrane** - a tissue called the synovial membrane lines the joint and seals it into a joint capsule. The synovial membrane secretes synovial fluid (a clear, sticky fluid) around the joint to lubricate it.
- **ligaments** - strong ligaments (tough, elastic bands of connective tissue) surround the joint to give support and limit the joint's movement. It joins bones to each other and gives the joint stability.
- **tendons** - tendons (another type of tough connective tissue) on each side of a joint attach muscles to bones. The tendons are inelastic.

Muscles work in pairs. As one contracts the other in the pair relaxes. One muscle produces the opposite movement from the other muscle, therefore, the pairs are called **antagonistic pairs**.

Muscles which cause a joint to extend are called **extensors**, muscles which cause a limb to retract, bend or fold are called **flexors**. **Flexion** is the movement that reduces the angle of a joint. **Extension** increases the angle of a joint and moves the bones away from each other.

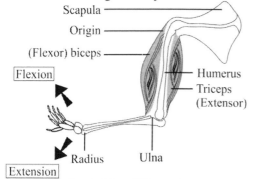

Fig.1.12 – Elbow joint

The biceps are attached to the scapula at one end and to the radius at the other end. So, when the bicep contracts the arm is bent at the elbow. This is called as flexing and the bicep is the flexor.

To straighten the arm or extend the arm, the triceps contract and pull the ulna. So, the triceps act as extensors.

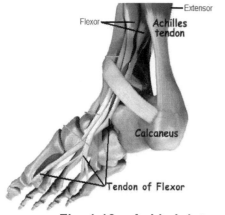

Fig. 1.13 – Ankle joint

The tendons join muscles to bones. The inelastic nature of tendons ensure that the bones are pulled when the muscles contract. This aids the movement of bones about a joint.

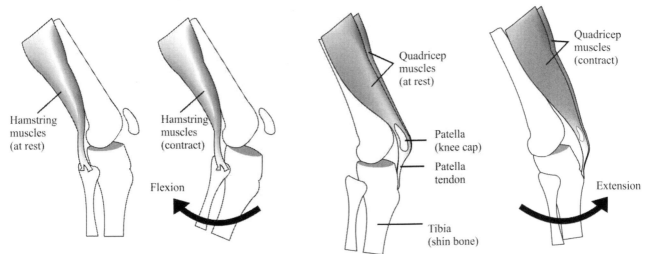

Fig. 1.14 – Extension and flexion of muscles

Structure	Role
Bone	Acts as a rigid lever
Muscles	Effort or force to move the bones of the lever system.
Tendons	Attach muscles to bones.
Ligaments	Stabilise the skeleton at joints. Joins bones to each other.
Nerves	Stimulate and coordinate muscle contraction.

CHAPTER TWO
CELLULAR RESPIRATION

Learning outcomes: by the end of this chapter you should be able to
Edexcel Syllabus Spec 5: *Describe the overall reaction of aerobic respiration as splitting of the respiratory substrate (eg glucose) to release carbon dioxide as a waste product and reuniting of hydrogen with atmospheric oxygen with the release of a large amount of energy.*
Edexcel Syllabus Spec 6: *Describe how to investigate rate of respiration practically.*
Edexcel Syllabus Spec 7: *Recall how phosphorylation of ADP requires energy and how hydrolysis of ATP provides an accessible supply of energy for biological processes.*
Edexcel Syllabus Spec 8: *Describe the roles of glycolysis in aerobic and anaerobic respiration, including the phosphorylation of hexoses, the production of ATP, reduced coenzyme and pyruvate acid (details of intermediate stages and compounds are not required).*
Edexcel Syllabus Spec 9: *Describe the role of the Krebs cycle in the complete oxidation of glucose and formation of carbon dioxide (CO2), ATP, reduced NAD and reduced FAD (names of other compounds are not required) and that respiration is a many-stepped process with each step controlled and catalysed by a specific intracellular enzyme.*
Edexcel Syllabus Spec 10: *Describe the synthesis of ATP by oxidative phosphorylation associated with the electron transport chain in mitochondria, including the role of chemiosmosis and ATPase.*
Edexcel Syllabus Spec 11: *Explain the fate of lactate after a period of anaerobic respiration in animals.*

Cellular respiration is an essential process that occurs in all living cells. It uses glucose to produce ATP, which provides energy for metabolic reactions.
The simplified equation for respiration is given below.

$$C_6H_{12}O_6 + 6O_2 \longrightarrow 6H_2O + 6CO_2 + 38\ ATP\ (energy)$$

However, respiration occurs through a series of reactions, namely glycolysis, link reaction, Kreb's cycle and oxidative phosphorylation. These reactions are referred to as metabolic pathways because every step in the reaction is controlled by a different enzyme.

ATP: energy to drive metabolic reactions. ATP is made up of a nitrogenous base (Adenine), a ribose sugar and three phosphate groups. The covalent bond linking the second and third phosphate group is unstable, and is easily broken by **hydrolysis**. When this bond is broken a phosphate group is removed, and **ATP becomes ADP**. Since **energy is released** during this reaction, it is called an **exergonic** reaction. The breakdown of ATP releases energy to do work. A contracting muscle cell requires about two million ATP molecules per second. ATP is **not** an energy storage molecule, it is used up within one minute after it is produced. All the ATP in our body will be enough to keep us alive only for a few minutes. ATP must be generated continuously to provide energy for metabolic reactions to continue.

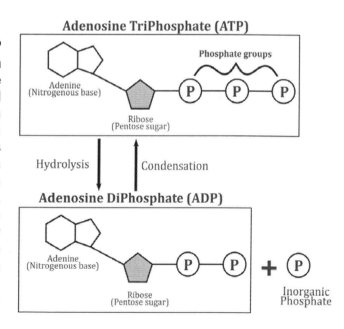

Fig. 2.1 – Condensation and Hydrolysis of ATP

ATP can be resynthesised from ADP and Pi (inorganic phosphate) by **condensation reactions.** The energy for this reaction comes from **respiratory substrates**. The ATP cycle shows the relationship between ATP, ADP and respiration. The condensation of ADP to ATP is an energy-consuming process, **Energy – consuming** processes are called **endergonic.**

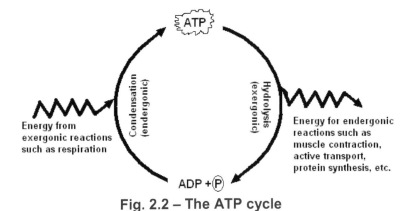

Fig. 2.2 – The ATP cycle

Glycolysis and Krebs cycle

Glycolysis, link reaction and Krebs cycle are metabolic pathways which involve many **enzyme controlled reactions**. The detailed pathway is given below. It is not necessary to memorise all the steps or names of intermediate compounds. Refer to fig. 2.5 for a summary of the process.

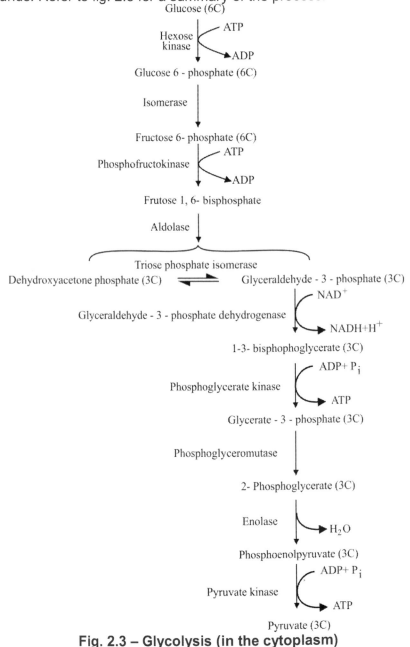

Fig. 2.3 – Glycolysis (in the cytoplasm)

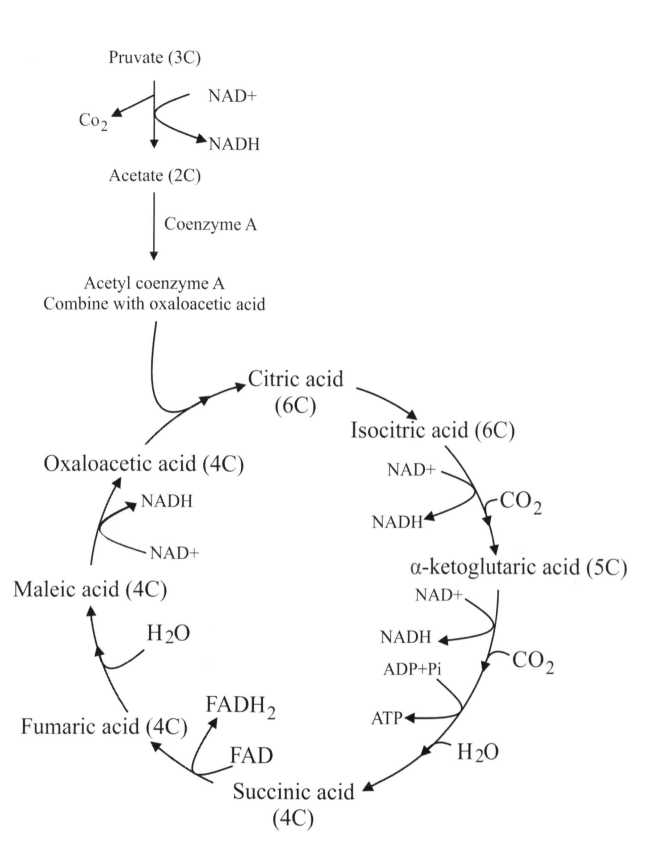

Pruvate (3C)

NAD+

Co_2

NADH

Acetate (2C)

Coenzyme A

Acetyl coenzyme A
Combine with oxaloacetic acid

Citric acid
(6C)

Isocitric acid (6C)

Oxaloacetic acid (4C)

NAD+

NADH

CO_2

NADH

NAD+

α-ketoglutaric acid (5C)

NAD+

Maleic acid (4C)

NADH

H_2O

ADP+Pi

CO_2

ATP

FADH$_2$

H_2O

Fumaric acid (4C)

FAD

Succinic acid
(4C)

**Fig. 2.4 – Link reaction and Krebs Cycle
(in the matrix of mitochondria)**

The chart below shows a summary of the events that occur during the glycolytic pathway and Kreb's cycle.

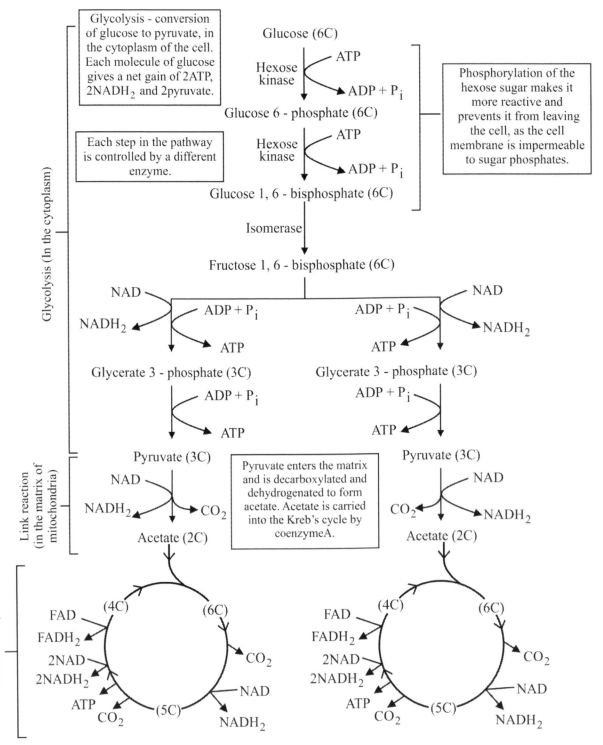

Glycolysis - conversion of glucose to pyruvate, in the cytoplasm of the cell. Each molecule of glucose gives a net gain of 2ATP, $2NADH_2$ and 2pyruvate.

Each step in the pathway is controlled by a different enzyme.

Glucose (6C)

Hexose kinase

ATP
$ADP + P_i$

Glucose 6 - phosphate (6C)

Hexose kinase

ATP
$ADP + P_i$

Glucose 1, 6 - bisphosphate (6C)

Phosphorylation of the hexose sugar makes it more reactive and prevents it from leaving the cell, as the cell membrane is impermeable to sugar phosphates.

Isomerase

Fructose 1, 6 - bisphosphate (6C)

Glycolysis (In the cytoplasm)

NAD
$NADH_2$
$ADP + P_i$
ATP

NAD
$ADP + P_i$
ATP
$NADH_2$

Glycerate 3 - phosphate (3C)

Glycerate 3 - phosphate (3C)

$ADP + P_i$
ATP

$ADP + P_i$
ATP

Pyruvate (3C)

Pyruvate (3C)

Link reaction (in the matrix of mitochondria)

NAD
$NADH_2$
CO_2

Pyruvate enters the matrix and is decarboxylated and dehydrogenated to form acetate. Acetate is carried into the Kreb's cycle by coenzymeA.

NAD
CO_2
$NADH_2$

Acetate (2C)

Acetate (2C)

Kreb's cycle (in the matrix of mitochondria)

FAD
$FADH_2$
2NAD
$2NADH_2$
ATP
CO_2
(4C)
(6C)
(5C)
CO_2
NAD
$NADH_2$

FAD
$FADH_2$
2NAD
$2NADH_2$
ATP
CO_2
(4C)
(6C)
(5C)
CO_2
NAD
$NADH_2$

The acetate (2C) combines with a 4C compound (oxaloacetate) to form a 6C compound (citrate). Citrate is decarboxylated and oxidised (dehydrogenated) to form a 5C compound (a ketoglutarate), CO_2 and $NADH_2$. The 5C compound is the decarboxlated, oxidised and dephosphorylated through a series of reactions, involving many intermediate compounds and enzymes, to regenerate oxaloacetate (4C). The net result of the kreb's cycle for each glucose molecule is $4CO_2$, $6NADH_2$, $FADH_2$, and 2ATP.

Fig. 2.5 – Summary of Glycolysis, link reaction and Krebs cycle

Oxidative phosphorylation: Phosphorylation of ADP to ATP in the presence of oxygen.
The diagram below summarises the events that occur during oxidative phosphorylation.

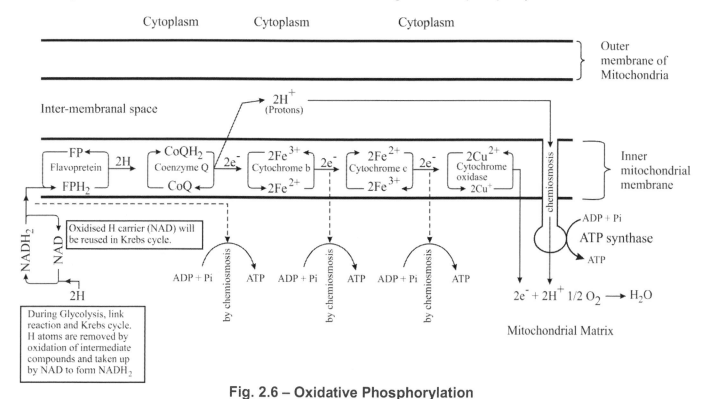

Fig. 2.6 – Oxidative Phosphorylation

➤ $NADH_2$ from glycolysis, link reaction and Krebs cycle gets oxidised by LOSING hydrogen to the electron transport chain.

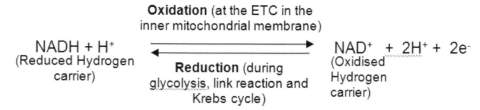

➤ The components of the electron transport chain split hydrogen into electrons and hydrogen ions. The electrons pass down the chain through a series of redox reactions, controlled by oxidoreductase or dehydrogenase enzymes.

➤ As the electrons are passed from one component to another, hydrogen ions are pumped from the matrix, into the inter-membranal space.

➤ This causes a high concentration of hydrogen ions in the inter-membranal space.

⭹ ATP is generated from free energy released when H^+ ions move back into mitochondrial matrix, through stalked particles, which contain the enzyme ATP synthase or ATPase. These stalked particles are called chemiosmotic channels and the movement of hydrogen ions into the matrix is called chemiosmosis.

⭹ Oxygen is used to absorb electrons from the electron transport chain and combines these electrons with H^+ ions to form water. This ensures that the Electron Transport Chain (ETC) continues to function and NAD^+ (oxidized Hydrogen carriers) can be regenerated for aerobic respiration to continue.

$$4H^+ + 4e^- + O_2 \longrightarrow 2H_2O$$

Note: During oxidative phosphorylation, one $NADH_2$ produces 3ATP and one $FADH_2$ produces 2ATP, as it joins the chain later and pumps in less hydrogen ions to the inter-membranal space.

Yield of ATP during aerobic respiration	
Glycolysis (ATP synthesized by substrate level phosphorylation)	2ATP
Krebs cycle (ATP synthesized by substrate level phosphorylation)	2ATP
Glycolysis (Each $NADH_2$ from glycolysis will yield Three ATP during oxidative phosphorylation.)	2 $NADH_2$ X 3ATP = 6ATP
Link reaction (Each $NADH_2$ from link reaction will yield Three ATP during oxidative phosphorylation.)	2 $NADH_2$ X 3ATP = 6ATP
Krebs cycle (Each $NADH_2$ from Krebs cycle will yield Three ATP during oxidative phosphorylation.)	6 $NADH_2$ X 3ATP = 18ATP
Krebs cycle (Each $FADH_2$ from Krebs cycle will yield two ATP during oxidative phosphorylation as it enters the electron transport chain, later than $NADH_2$, between FP and CoQ)	2 $FADH_2$ X 2ATP = 4ATP
Total ATP yield from one molecule of glucose during aerobic respiration =	**38 ATP**

Reduction:	Oxidation:
<ul><li>Loss of oxygen;</li><li>Gain of hydrogen;</li><li>Gain of electrons.</li></ul>	<ul><li>Gain of oxygen;</li><li>Loss of hydrogen;</li><li>Loss of electrons.</li></ul>

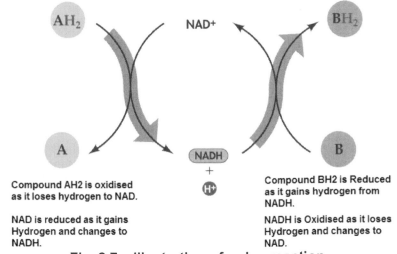

Fig. 2.7 – illustration of redox reaction

The reaction in fig.2.7 shows the effect of **dehydrogenase enzymes** which remove hydrogen from the intermediate compounds in the respiratory pathway.

Likewise, every transfer of electron from one carrier to the next involves a redox reaction, which is controlled by **oxidoreductase enzymes**. Cytochrome oxidase is an example of an oxidoreductase enzyme which catalyses the transfer of hydrogen to oxygen at the end of the electron transport chain, to form water. Cyanide inhibits this enzyme and causes the Electron Transport Chain to stop functioning, leading to death.

Anaerobic respiration in animals

During anaerobic conditions Oxygen is not available to accept electrons from the electron transfer chain. This causes the electron transfer chain to stop functioning. Oxidized hydrogen carriers (NAD^+) cannot be regenerated. Without Oxidized hydrogen carriers (NAD^+), Krebs cycle cannot operate. Thus, pyruvate cannot be further broken down by respiration.

However, an alternate pathway in the **cytoplasm** can generate enough Oxidized hydrogen carriers (NAD^+) during anaerobic conditions. These Oxidized hydrogen carriers (NAD^+) can be reused for glycolysis to continue. So, during anaerobic respiration, glycolysis is the only source of ATP production. Hence, a single glucose molecule can yield only 2ATP during anaerobic respiration, as shown in fig. 2.8.

Anaerobic respiration in muscle cells

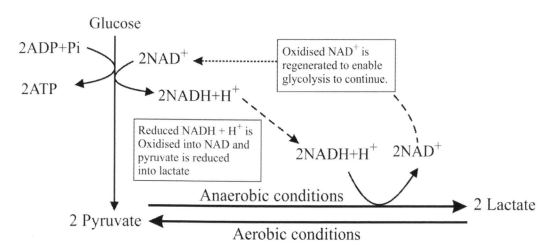

Fig. 2.8 – alternate pathway regenrates NAD for glycolysis to continue

When exercise stops, the levels of lactate in the blood remain raised. The lactate changes into lactic acid and lowers the pH in the muscles causing muscle cramps, due to denaturation of respiratory enzymes and muscle proteins. The denaturation of respiratory enzymes leads to lack of ATP and muscle cramps.

The lactate must be oxidised back to pyruvate. The pyruvate can then undergo aerobic respiration. It takes oxygen to oxidise the accumulated lactate. The oxygen needed to oxidize the lactate is called as the oxygen debt. This is why you continue to breathe deeply for some time after exercise has finished.

The chart below shows how the amount of oxygen used by the body changes over time. At the beginning the body works anaerobically leaving an oxygen deficit. Over time the oxygen consumption levels out to a steady state. After exercise the oxygen is paid back (oxygen debt). Notice the area of oxygen debt (X) is greater than the area of oxygen deficit (Y). This is because of EPOC (Excess post-exercise oxygen consumption). This involves the use of oxygen in some of the following processes:

- Myoglobin in the muscles needs to be reoxygenated.
- Transport of lactate to the liver and its subsequent conversion into glucose by gluconeogenesis.

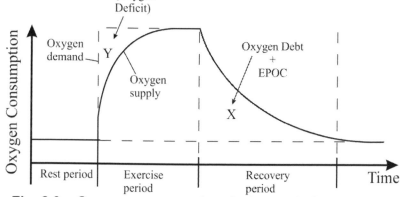

Fig. 2.9 – Oxygen consumption during and after exercise

Investigating the rate of respiration practically by using a respirometer

Respiration is a metabolic activity which occurs in all living cells. It may be aerobic or anaerobic. The rate of respiration can be measured either by measuring the rate of oxygen used or the rate of CO_2 evolved. The diagram below shows a respirometer used to measure the rate of oxygen used by living tissue.

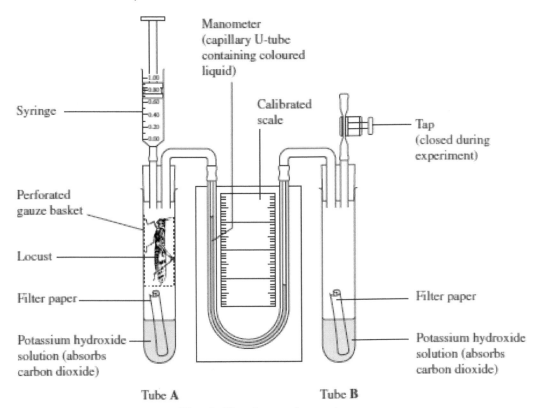

Fig. 2.10 – A respirometer

Step one: Equilibration of pressure and temperature

Place tubes **A** and **B** in a water bath at 30 °C for 10 minutes. Remove the syringe tube **A** and open the tap attached to tube **B**. **This allows time for the pressure inside the tubes to neutralise. The pressure in the tubes changes due expansion of air and the glass tubes.**

Step two: Setting the starting point

The syringe is then attached to tube **A** and the tap on tube **B** is left open. The syringe is then used to set the liquid level in both tubes to be the same.

The tap in tube B is then closed. The apparatus is now ready for use.

Step three: measuring the rate of respiration.

The reading on the manometer scale is then recorded. The rate of respiration is given by the following expression.

$$\text{Rate of respiration} = (\pi\, r^2\, l)\, /\, (t \times m)$$
Or
$$\text{Rate of respiration} = (\text{Volume of oxygen used})\, /\, (\text{time} \times \text{mass})$$

Where,

π – 3.14

r – diameter of manometer tube

l – length travelled by the liquid column

t – refers to duration for which the liquid column moves.

m – refers to mass of living tissue.

$(\pi\, r^2\, l)$ – volume of oxygen used.

CHAPTER THREE
ELECTROCARDIOGRAMS (ECGs)

Learning outcomes: by the end of this chapter you should be able to

Edexcel Syllabus Spec 12: Understand that cardiac muscle is myogenic and describe the normal electrical activity of the heart, including the roles of the sinoatrial node (SAN), the atrioventricular node (AVN) and the bundle of His, and how the use of electrocardiograms (ECGs) can aid the diagnosis of cardiovascular disease (CVD) and other heart conditions.

The electrocardiogram

The electrocardiogram (ECG) records the electrical activity of the heart. Normally a patient will have 12 electrodes attached to their body. Each lead used with an ECG looks at the heart from a different angle. In the explanation that follows, only a single trace will be used. The positions of six leads are shown in the diagram below. The other leads are placed on the limbs.

ECG Lead Positions

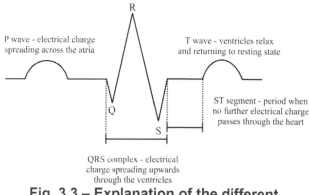

Fig. 3.1 – ECG leads

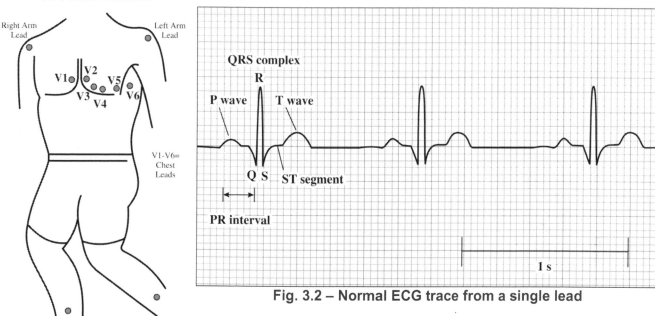

Fig. 3.2 – Normal ECG trace from a single lead

P wave - electrical charge spreading across the atria

T wave - ventricles relax and returning to resting state

ST segment - period when no further electrical charge passes through the heart

QRS complex - electrical charge spreading upwards through the ventricles

Fig. 3.3 – Explanation of the different segments of an ECG.

> The **P wave** is the result of a wave of electrical charge spreading across the atria. When it reaches the atrioventricular node at the base of the right atrium there is a slight delay shown by the time between the end of the P wave and the start of the QRS complex; this allows the atria to finish contracting before the ventricle contracts. The P wave is not as elevated as the QRS complex as the atria have thinner wall compared to the ventricles.

> The **QRS complex** is due to electrical charge spreading upwards through the ventricles.

> The **ST segment** is the short period of time when no further electrical impulse can be passed through the heart muscle.

> The **T wave** is the period when the ventricles are relaxing and return to their resting state.

Using the ECG trace to measure heart rate

The ECG trace can be used to measure heart rate. The squared paper passing through an ECG machine moves at a steady 25 mm per second (5 large squares per second).

This means that 300 of the large squares will pass through in 1 minute (60 seconds x 5 squares = 300 large squares). One large square is equivalent to 0.2 seconds.

Heart rate can be determined by finding the average number of large squares between two QRS complexes. This value is divided by 300 to give the heart rate.

For example the normal ECG trace shown below has Five large squares between the QRS complexes, so the heart rate is: 300 ÷ 5 = 60 beats per minute.

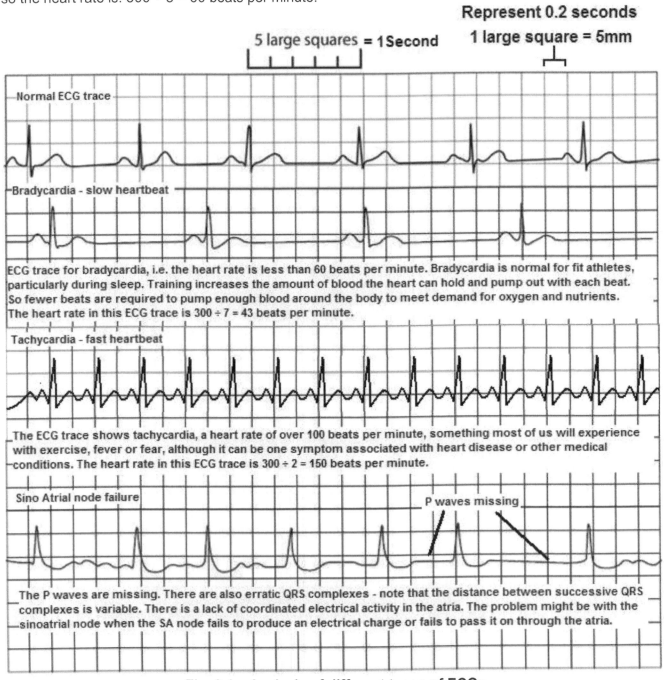

Represent 0.2 seconds
1 large square = 5mm

5 large squares = 1 Second

Normal ECG trace

Bradycardia - slow heartbeat

ECG trace for bradycardia, i.e. the heart rate is less than 60 beats per minute. Bradycardia is normal for fit athletes, particularly during sleep. Training increases the amount of blood the heart can hold and pump out with each beat. So fewer beats are required to pump enough blood around the body to meet demand for oxygen and nutrients. The heart rate in this ECG trace is 300 ÷ 7 = 43 beats per minute.

Tachycardia - fast heartbeat

The ECG trace shows tachycardia, a heart rate of over 100 beats per minute, something most of us will experience with exercise, fever or fear, although it can be one symptom associated with heart disease or other medical conditions. The heart rate in this ECG trace is 300 ÷ 2 = 150 beats per minute.

Sino Atrial node failure

P waves missing

The P waves are missing. There are also erratic QRS complexes - note that the distance between successive QRS complexes is variable. There is a lack of coordinated electrical activity in the atria. The problem might be with the sinoatrial node when the SA node fails to produce an electrical charge or fails to pass it on through the atria.

Fig. 3.4 – Analysis of different types of ECGs

Note: Bradycardia and tachycardia are referred to as **arrhythmia**, because the heartbeats are not to the normal rhythm.

A patient who suffers chest pain, which may be due to development of atherosclerosis in the coronary arteries giving rise to angina, will show abnormal-shaped waves on their ECG. During the period of pain the ST segment lies below the normal position and there may be inversion of the T wave.

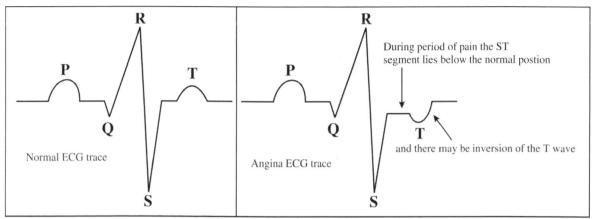

Fig. 3.5 – T wave inversion

In patients with pulmonary hypertension, blood pressure in the lungs is much higher than normal. Their ECG trace has a large P wave. This is because the right atrium has to generate a higher pressure to get blood into ventricles because the right ventricle struggles to pump blood to the lungs due to the high pressure in lungs.

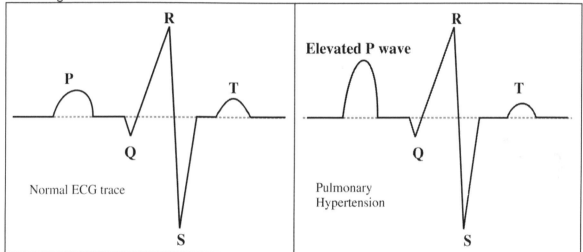

Fig. 3.4 – P wave elevation

Ventricullar fibrillation is a condition in which the heart's electrical activity becomes disordered. When this happens, the heart's lower (pumping) chambers contract in a rapid, unsynchronized way. (The ventricles "flutter" rather than beat.) The heart pumps little or no blood.

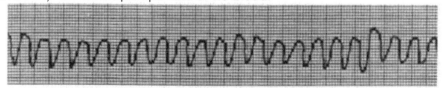

Fig. 3.4 – Ventricular fibrillation

Ventricular fibrillation is very serious. Collapse and sudden cardiac death will follow in minutes unless medical help is provided immediately. This requires shocking the heart with a device called a defibrillator. One effective way to correct life-threatening rhythms is by using an electronic device called an implantable cardioverter-defibrillator. This device shocks the heart to normalize the heartbeat if the heart's own electrical signals become disordered

CHAPTER FOUR
REGULATION OF HEARTBEAT RATE
AND
BREATHING RATE

Learning outcomes: by the end of this chapter you should be able to
Edexcel Syllabus Spec 13: *Explain how variations in ventilation and cardiac output enable rapid delivery of oxygen to tissues and the removal of carbon dioxide from them, including how the heart rate and ventilation rate are controlled and the roles of the cardiovascular control centre and the ventilation centre.*

When an average individual is at rest, the normal heartbeat rate is about 72 beats per minute and the breathing rate is about 14 breaths per minute. The cardiac output and ventilation rate is sufficient enough to supply the muscles with the oxygen they need and to carry away the Carbon dioxide at the required rate. However, as a person starts to exercise there is a greater need for oxygen to be supplied at a greater rate. The carbon dioxide must also be removed more rapidly. This is achieved by an increase in the cardiac output and the ventilation rate.

During exercise: the sympathetic control speeds up heartbeat rate and Frank–Starling effect increases the stroke volume.
Stroke volume is the volume of blood pumped by the ventricle in one contraction.
Heartbeat rate is the number of heartbeats in one minute.
Cardiac output (CO) is the volume of blood pumped by each side of the heart in one minute.

Cardiac output = heart rate x stroke volume

The mechanism by which the cardiac output increases during exercise is shown in Fig. 4.1

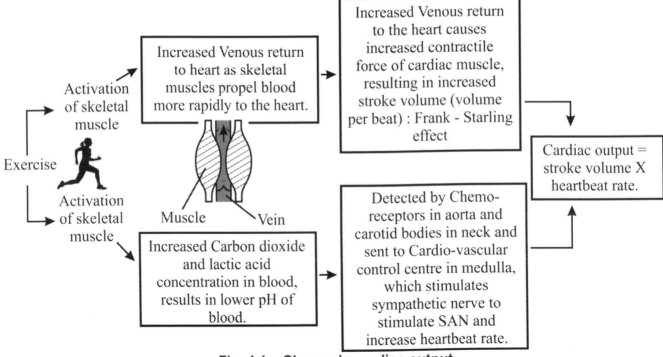

Fig. 4.1 – Change in cardiac output

The increased cardiac output ensures that
- oxygen is supplied rapidly to muscles to meet the excess demand during exercise.
- carbon dioxide and lactate is removed rapidly from muscles (otherwise pH could become too low and inhibit respiratory enzymes).
- blood circulates rapidly through the lungs to ensure rapid oxygenation and removal of carbon dioxide.

After exercise, the influence of sympathetic nerves on the SAN decreases, as the carbon dioxide and lactic acid are no longer produced rapidly. This restores normal rhythm of the SAN (myogenic).

The heartbeat rate and stroke volume (pressure) are prevented from spiralling out of control by stretch receptors in the aorta and carotid artery. These receptors detect high blood pressure and send impulses to the cardiovascular control centre in the medulla. The inhibitory region, of the cardiovascular centre will send impulses to the Sino-Atrial Node (SAN), through the parasympathetic (vagus) nerves to slow down the heartbeat rate and decrease the stroke volume.

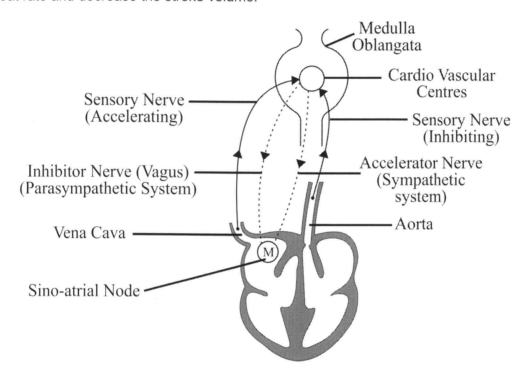

Fig. 4.2 – Sympathetic and parasympathetic nerves

Control of breathing rate and depth (ventilation rate)
Breathing is controlled by **respiratory centers** in the hindbrain - medulla oblongata, which controls the rate and depth of breathing. The centre is referred to as the **medullary rhythmicity centre.** It controls the basic regular rhythm of breathing. However, the rate and depth of breathing can be controlled by stimulation of the **inspiratory and expiratory centres, from the chemoreceptors** in the medulla, carotid bodies and aorta.

Inspiratory centre: increases the rate and depth of inspiration during exercise, when CO_2 concentration of blood increases, pH decreases or oxygen concentration decreases.
Expiratory centre: stimulates the rate and depth of expiration.
Receptors: Stretch receptors in bronchial tree, chemoreceptors in blood vessels and medulla oblongata of the brain.
Effectors: Intercostal muscles, diaphragm.

The basic functioning of the respiratory centre is illustrated in fig. 4.3.

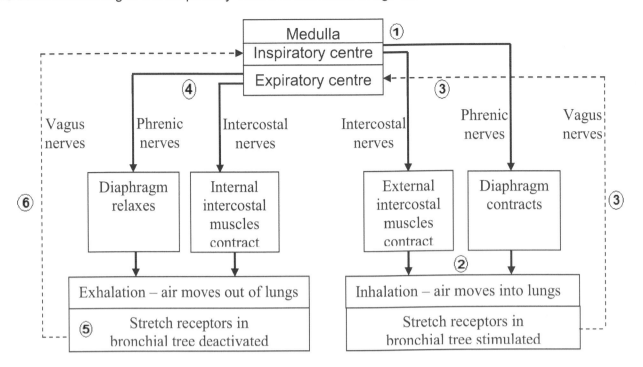

Fig. 4.3 – Control of breathing rate and volume

1. External intercostal muscles and diaphragm are stimulated by the inspiratory centre to contract.
2. Lungs inflate and stretch receptors in bronchial tree are stimulated.
3. Vagus nerve stimulates the expiratory centre to stop inhalation and begin exhalation.
4. Internal intercostal muscles contract and diaphragm relaxes due to stimulation by the expiratory centre.
5. Lungs deflate and stretch receptors in bronchial tree are deactivated.
6. Vagus nerve stimulates the inspiratory centre to stop exhalation and begin inhalation.

Changes in breathing rate during exercise

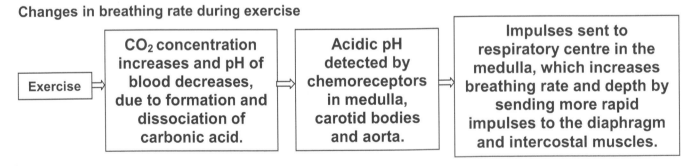

Fig. 4.4 – Change in Breathing rate during exercise

After exercise, the rate of breathing slows down as the carbon dioxide in the blood decreases. The events which lead to the decrease are stated below.

1. There is less carbon dioxide dissolved in the blood plasma so less carbonic acid forms and dissociates.
2. The pH of the blood rises.
3. Chemoreceptors in the ventilation centre of the medulla oblongata detect the change in pH.
4. Impulses are sent to other parts of the ventilation centre.
5. Fewer impulses are sent from the ventilation centre to stimulate the muscles involved in breathing.
6. Fewer and weaker contractions of the external intercostal muscles and diaphragm muscles decrease the rate and depth of breathing.

CHAPTER FIVE
SPIROMETRY

Learning outcomes: by the end of this chapter you should be able to
Edexcel Syllabus Spec 14: Describe how to investigate the effects of exercise on tidal volume and breathing rate using data from spirometer traces.

Lung volumes

Inspiratory reserve volume (IRV): the maximum volume of air that can be forcibly inhaled after a normal inhalation.

Expiratory reserve volume (ERV): the maximum volume of air that can be forcibly exhaled after a normal exhalation.

Tidal volume (TV): the volume of air breathed in or out of lungs per breath, during normal breathing at rest. It is about $0.5dm^3$ at rest, but varies with each individual. It increases during exercise.

Vital capacity (VC): the maximum volume of air that can be forcibly expired after a maximal intake of air. It varies between $3\ dm^3$ to $6\ dm^3$ depending on size and fitness of the person.

VC = IRV + TV + ERV

Residual Volume (RV): it is the volume of air that remains in the lungs even after a forced expiration. It prevents the lungs from collapsing after expiration.

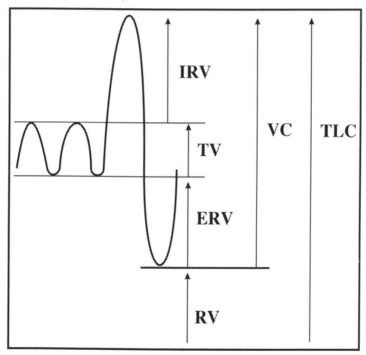

Fig. 5.1 – Lung volumes

The ventilation rate of an individual and the changes during exercise is shown below.

breathing rate x tidal volume = ventilation rate

Both controlled via the nerves from the respiratory centre.

	breathing rate (breaths/min)	tidal volume (cm³ / breath)	ventilation rate (cm³ / min)
at rest	12	500	6 000
at exercise	18	1000	18 000

Lung volumes can be measured by using a spirometer. The fig. 5.2 shows the components of a spirometer. It is not compulsory to learn the details of the working, as the specification only requires you to interpret spirometer traces before and after exercise.

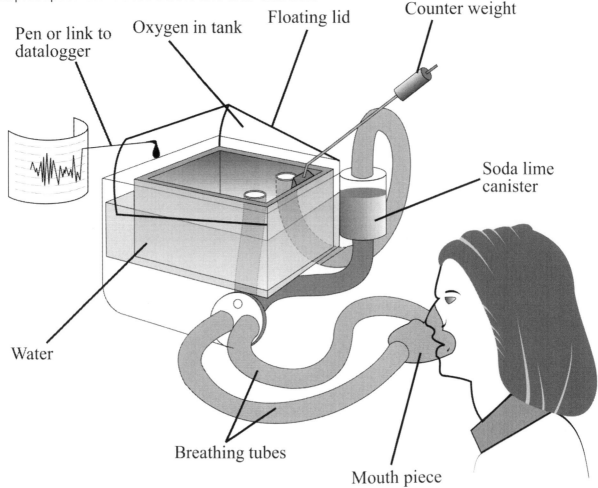

Fig. 5.2 – A spirometer

Working of spirometer

When the air tank is filled with oxygen and the water tank filled with water, you can breathe the oxygen through the breathing tubes. The lid of the air tank goes up and down as you breathe in and out, making a trace on the chart recorder. The speed of the chart recorder can be set, so the relationship between distance and time can be calculated on the trace. The volume readings can be **calibrated** by making marks on the chart with the pen before and after a known volume of oxygen is added to the air tank. Breathing rate is calculated by counting the number of breaths in a given time. The volume of air breathed in and out is calculated from the vertical movements of the trace.

Noseclip

The nose clip makes sure that **all breathing is only through the air tubes of the spirometer**, so that the trace reflects the true volume of gas breathed in and out.

Sodalime

The soda lime absorbs CO_2 in the exhaled air. This ensures that the CO_2 levels in the inhaled air do not change during the experiment, making uptake of oxygen easier to determine. The removal of CO_2 is also necessary for safety reasons.

Example one

The spirometer chart in fig.5.3 shows two traces; one at rest (lower trace) and one after exercise (upper trace).

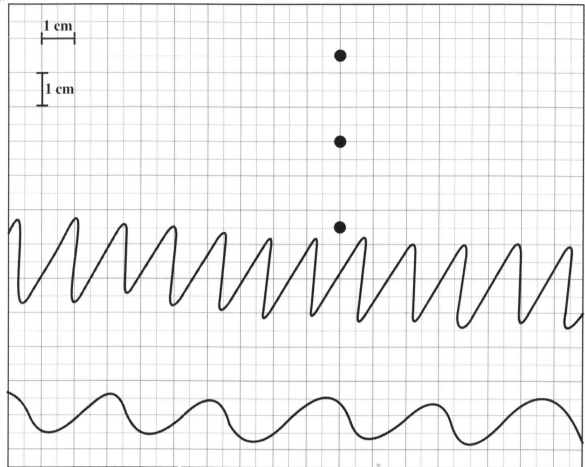

Fig. 5.3 – Spirometer trace

The three dots above the upper trace are the calibration dots: the first dot, the lowest of the three, is the baseline level recorded before any oxygen is added into the air tank, the second dot is after adding 1 dm³ oxygen to the chamber within the spirometer, and the third dot is after adding another 1 dm³ oxygen. The chart recorder was set at **0.5 cm s⁻¹**.

(NB 1 dm³ is the same as 1 litre; 1 dm³ = 1000 cm³.)

Breathing rate measurement

Breathing rate before exercise: 4 breaths in 26 or 27 seconds; 4 × 60/26 (or 27) = 9 breaths per minute.
Breathing rate after exercise: 11 breaths in 34 seconds; 11 × 60/34 = 19 breaths per minute.
Breathing rate increases by about 10 breaths per minute; It more than doubles (increasing by a factor of 2.1).

Tidal volume measurement

Tidal volume before exercise: average vertical height of trace = 1.3 cm.

From calibration 2.5 cm = 1 dm³. 1 dm³ ÷ 2.5 cm × 1.3 cm = 0.52 dm³.

Tidal volume after exercise: average vertical height of trace = 2.4 cm.
1 dm³ ÷ 2.5 cm × 2.4 cm = 0.96 dm³.

Tidal volume increases by 0.44 dm³ after exercise; it nearly doubles (increasing by a factor of 1.8).

Example two

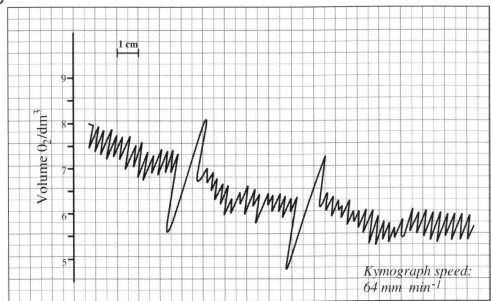

Fig. 5.4 – Spirometer trace for spirometer used with soda lime and filled to 9 dm^3 with oxygen.

Tidal volume – between 0.5 and 0.8 dm^3. Note: It is advisable to take an average over several breaths.
Vital capacity = 2.55 dm^3 averaged over the two breaths.
Breathing rate = between 20 and 22 breaths per minute (depending on the section of the graph used).
Minute ventilation = 21 breaths per minute × 0.65 dm^3 = 14 dm^3 min^{-1}.

Measuring oxygen consumption

Each time we take a breath, some oxygen is absorbed from the air in the lungs into our blood. An equal volume of carbon dioxide is released back into the lungs from the blood.

When we use the spirometer, each returning breath passes through soda lime, which absorbs the carbon dioxide, so less gas is breathed back into the spirometer chamber than was breathed in.

If we breathe into and out of the spirometer for (say) 1 minute, a steady fall in the spirometer trace can be seen. The gradient of the fall is a measure of the rate of oxygen absorption by the blood, and so is a measure of the rate of respiration by the body.

The rate of oxygen consumption determined from the trace in the figure above is approximately 1.0 dm^3 min^{-1} for the first 30 seconds.

If the person had been exercising, the rate of oxygen consumption would be higher, so the slope would be steeper. The trace would also show that the subject was breathing faster and more deeply.

Example three

The changes in breathing rate and volume reflect increased gas exchange during exercise due to the increased demand for energy in skeletal muscles. The supply of oxygen to the tissues is increased where oxygen is needed for aerobic respiration. More energy is made available for muscle contraction during exercise. The graph should show that breaths become deeper and more frequent **after exercise**, with a greater rate of oxygen consumption.

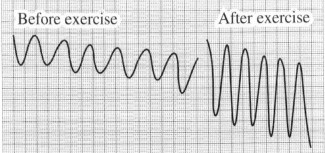

Fig. 5.5 – Trace before and after exercise

CHAPTER SIX
FEEDBACK MECHANISMS

Negative feedback

Homeostasis (a Greek term meaning same state), is the maintenance of constant conditions in the internal environment of the body despite large swings in the external environment. Functions such as blood pressure, body <u>temperature</u>, respiration rate, and blood glucose levels are maintained within a range of normal values around a set point despite constantly changing external conditions.

The body's homeostatically cultivated systems are maintained by negative feedback mechanisms, sometimes called negative feedback loops. In negative feedback, any change or deviation from the normal range of function is opposed, or resisted. The change or deviation in the controlled value initiates responses that bring the function of the <u>organ</u> or structure **back to within the normal range.**

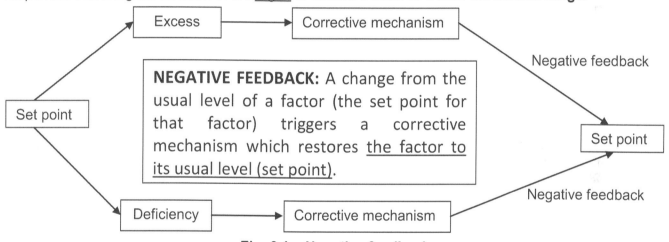

Fig. 6.1 – Negative feedback

The positive feedback mechanism is shown below. It is not specified in the syllabus, but helps to understand negative feedback better.

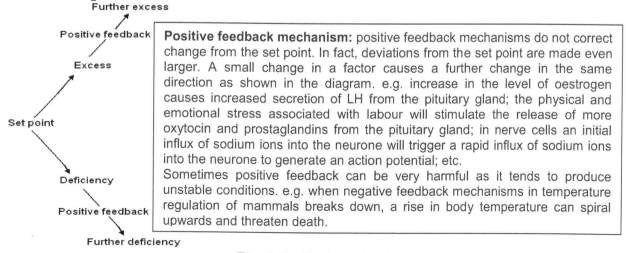

Fig. 6.2 – Positive feedback

Negative feedback and thermoregulation

```
                    ┌─────────────────────────────┐
                    │  Normal body temperature     │
                    │           37⁰C               │
                    └─────────────────────────────┘
```

Normal body temperature
37^{0}C

Increase in temperature
- Exercise / high metabolism
- Warm clothes
- Heat gain from surrounding
  / warm surrounding

Decrease in temperature
- Physical inactivity / low metabolism
- Lack of clothing
- Heat loss to surrounding
  / cold surrounding

Increase in temperature is detected by temperature sensitive **receptors in the hypothalamus and skin**

Decrease in temperature is detected by temperature sensitive **receptors in the hypothalamus and skin**

Impulses sent to **temperature control centre in the hypothalamus (Thermo-regulatory centre)**

Impulses sent to **temperature control centre in the hypothalamus (Thermo-regulatory centre)**

Thermoregulatory centre brings about following changes
- **Vasodilation** – smooth muscles of arterioles relax and more blood flows into capillaries below skin. This increases heat loss by radiation.
- **Hair erector muscles relax** – this results in the hair falling towards the skin and less air is trapped around the body. Heat loss by conduction increases.
- **Increased sweat production** – Evaporation of sweat from skin cools the body by removing heat.
- **Decreased metabolism** – the metabolic activity in liver and muscles is reduced. This helps to produce less heat.
- **Behavioral changes** – pouring water on the skin or turning on the fan helps to cool the body down.

Thermoregulatory centre brings about following changes
- **Vasoconstriction** – smooth muscles of arterioles constrict and less blood flows into capillaries below skin. This decreases heat loss by radiation.
- **Hair erector muscles contract** – this results in the hair becoming erect. Air is trapped around the body. Heat loss by conduction decreases (as air is an insulator).
- **Decreased sweat production** – Less evaporation of sweat from skin reduces heat loss.
- **Increased metabolism** – the metabolic activity in liver and muscles is increased. This helps to produce more heat.
- **Behavioral changes** – putting on warm clothes or turning on the heater helps to reduce heat loss.

Normal body temperature
37^{0}C

Fig. 6.3 – Thermoregulation

CHAPTER SEVEN
GENE INDUCTION

Learning outcomes: by the end of this chapter you should be able to
Edexcel Syllabus Spec 17: *Explain how genes can be switched on and off by DNA transcription factors including hormones.*

'Switching on a gene' – Gene induction or activation
Transcription of a gene is initiated by **RNA polymerase** and **transcription factors** binding to a **promoter region** (section of DNA upstream to a gene).

RNA Polymerase + Transcription factors = Transcription Initiation Complex

Some transcription factors are always present in all cells (example the transcription factors needed to switch on the genes for respiration or protein synthesis). Other transcription factors are only synthesised in certain cells at a particular stage of development, often in an inactive form, which is later activated by **signal proteins or regulator proteins**. Signal proteins may act directly by entering the cell (like steroid hormones) or indirectly through a **second messenger (cAMP)**.

The gene is **'switched on'** when **all the transcription factors**, in <u>their active form</u>, are present.

'Switching off a gene' – deactivation
Genes are switched off (not able to be transcribed) by the cell
- **protein repressor molecules** may attach to **the promoter region**, hence blocking the attachment sites for transcription factors.
- **protein repressor molecules** can attach to **the transcription factors** preventing them forming the transcription initiation complex.
- **signal proteins (Hormones)** acting as transcription factors may not be present.
- When the DNA is coiled around histones (Supercoiling), certain genes may be inactivated because its promoter may not be accessible to the transcription factors or RNA polymerase.

Every transcription initiation complex has a range of different factors with some acting as activators and others repressing expression. The exact order in which these factors bind is not known with certainty. In some experiments the transcription factors seem to bind in a specific order. In other cases, most of the factors are thought to first assemble with the polymerase, with this whole assembled complex then binding to the DNA in a single step.

The number of transcription factors found within an organism increases with the size of the genome. There are thought to be approximately 2600 proteins in humans that can bind to DNA. If most of these function as transcription factors then about 10% of our genes must code for transcription factors. This makes transcription factors the largest group of human proteins. The general transcription factors have been highly conserved in evolution, with the same factors occurring in different organisms from humans to simple single-cell organisms.

Mammalian hormones are chemical messengers carried by the blood from endocrine glands to all parts of the body. They affect specific cells, called **target cells**. Many hormones activate transcription factors in the target cells.
Chemically hormones are of three types:
- **Amines**: - Adrenaline / Thyroxine (**Lipid insoluble,** so cannot enter the cell).
- **Peptides/Proteins**: - Insulin / Glucagon (**Lipid insoluble,** so cannot enter the cell).
- **Steroids**: - Oestrogen / Testosterone (Steroids are **lipid soluble** and can enter the cell).

The mode of action for peptide hormones and steroid hormones is entirely different.

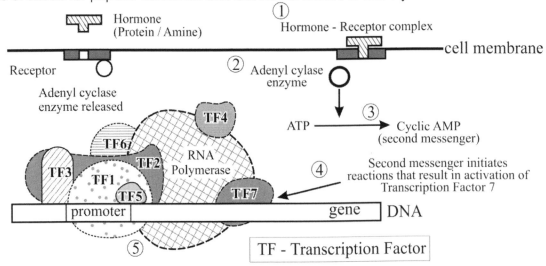

Fig. 7.1 – A peptide hormone

1. Hormone binds to specific receptor on the cell membrane.
2. Adenyl cyclase enzyme is released from the receptor and diffuses into the cytoplasm.
3. Adenyl cyclase converts ATP into cyclic AMP (Second messenger)
4. The second messenger initiates a series of reactions in the cell and activates Transcription Factor 7,in this case.
5. The activated transcription factor (TF7) now binds to the existing Transcription factors (TF1 to TF6, in this case) and completes the **Transcription Initiation Complex.** This activates RNA polymerase to become active and begin the process of transcription of the gene. The gene is now 'SWITCHED ON'.

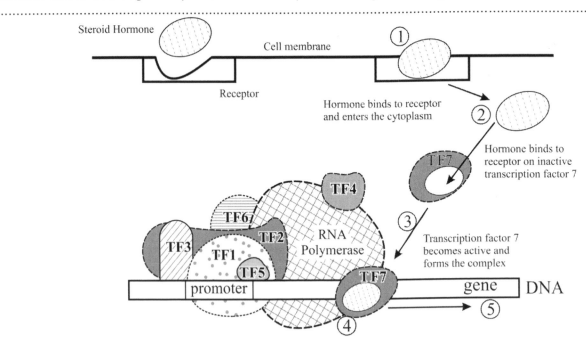

Fig. 7.2 – A steroid hormone

1. Steroid hormone binds to specific receptor on target cell and enters the cytoplasm, as the membrane is permeable to steroid hormones, like testosterone or oestrogen.
2. The steroid hormone will bind to a transcription factor in the cytoplasm.
3. The transcription factor and the hormone bind to other transcription factors (TF1 to TF6, in this case) and complete the formation of the **Transcription Initiation Complex.**
4. The formation of the **Transcription Initiation Complex** ensures that the gene is **'SWITCHED ON'.**
5. Transcription of the gene begins and mRNA is formed.

CHAPTER EIGHT
BENEFITS AND RISKS OF EXERCISE

Learning outcomes: by the end of this chapter you should be able to
Edexcel Syllabus Spec 18: Analyse and interpret data on possible disadvantages of exercising too much (wear and tear on joints, suppression of the immune system) and exercising too little (increased risk of obesity, coronary heart disease (CHD) and diabetes), recognising correlation and causal relationships.

Research tells us that regular moderate exercise is not only good for our muscles and hearts, but also for our immune systems. But too much of exercise could lead to tissue damage and immune suppression. Exercise can be our greatest protection against a host of illnesses and infections, but if over-done, we can leave ourselves defenseless from invaders normally caught by our immune systems. So get out there and work up a sweat and feel the burn, just don't overdo it.

Positive effects of moderate exercise include;

1. Increased BMR;	8. Increased bone density
2. Decreased blood pressure	9. Improved well being
3. Increased HDL	10. Decreased adrenaline levels
4. Decreased LDL	11. Less stress
5. Decreased risk of CHD	12. Moderate exercise increases levels of **Natural Killer** cells, which secrete apoptosis-inducing chemicals in response to non-specific viral or cancerous threat.
6. Maintaining healthy BMI	
7. Decreased risk of diabetes	

Disadvantages of exercising too much (over-training) include;

✓ **Chronic fatigue (tiredness) and poor athletic performance** – due to the body being exposed to constant stress and depletion of energy reserves. The body does not get enough time to repair damaged tissue and increases the risk of further tissue damage.

✓ **Increase in upper respiratory tract infections (URTI)** - Sore throats and flu-like symptoms, due to a **suppressed immune system** with a decrease in the number and activity of cells in the immune system. The immune system is suppressed due to stress hormones cortisol and adrenaline. These hormones decrease the number and activity of Natural Killer Cells, Phagoctyes, B lymphocytes & T lymphocytes. This decreases the immune response. Infection risk and immune function are related to the exercise workload, as shown in fig. 8.1.

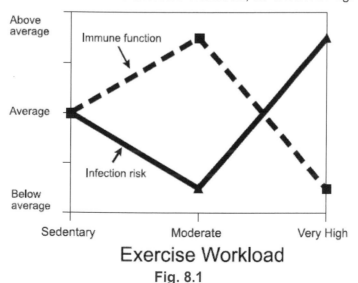

Exercise Workload

Fig. 8.1

✓ An **inflammatory response** in muscles due to damage to muscle fibres.

✓ **Increased wear and tear of joints**, which may require surgical repair. Damage to articular cartilage can lead to inflammation and a form of arthritis. The bursae (fluid sacs) which cushion the points of contact between bones, tendons and ligaments can swell up with extra fluid. As a result, they may push against other tissues in the joint, causing inflammation and tenderness.

✓ **Increased risk of cardiac failure,** specially in older untrained individuals. Over exercise results in increased adrenaline levels and increased stress. This increases the risk of cardiac arrest and stroke.

Disadvantages of exercising too little

➢ **Increased risk of weight gain** if energy input is high, leading to **obesity** (High BMI).

➢ **Increased risk of high blood pressure**, coronary heart disease and stroke.

➢ There will be no increase in the levels of blood HDL, nor reduction in LDL associated with exercise, which protect against coronary heart disease and stroke.

➢ Loss of sensitivity of cells to insulin which increases the likelihood of developing **type II diabetes**.

➢ No protection against loss of bone density and the development of osteoporosis.

➢ Greater risk of some cancers, due to a decrease in natural killer cells, which protect against cancerous tissue.

Correlation or causation: At low and moderate intensity of training there is a negative correlation between level of training and occurrence of Upper Respiratory Tract Infections (URTI). This means that as the level of exercise increases from low to moderate, the risk of URTI decreases.

As the intensity of exercise increases from moderate to high intensity, there is an increase in the risk of URTI. These variables show a positive correlation.

There is correlation between the two variable but this **does not necessarily mean that the change in one is responsible for the change in the other factor.**

Note **that a correlation can be considered as causal when all other factors which affect URTI are controlled. Factors such as, anxiety due to athletic events, infections acquired from others and diet must be ruled out before we conclude that changes in the risk of infections is CAUSED by exercise.**

There must be a clear and valid biological explanation to show how one factor causes the other to change for us to establish a causal relationship.

CHAPTER NINE
LAPROSCOPY, SPORT INJURIES AND ETHICS OF USING DRUGS

Learning outcomes: by the end of this chapter you should be able to

Edexcel Syllabus Spec 19: *Explain how medical technology, including the use of keyhole surgery and prostheses, is enabling those with injuries and disabilities to participate in sports, eg cruciate ligaments repair using keyhole surgery and knee joint replacement using prosthetics.*

Edexcel Syllabus Spec 20: *Outline two ethical positions relating to whether the use of performance-enhancing substances by athletes is acceptable.*

Knee
Normal anatomy

The knee is a complex joint made up of the femur, the tibia and fibula. A number of ligaments run between the femur and the tibia in the knee joint. The anterior cruciate ligament (ACL), the posterior cruciate ligament (PCL), and the collateral ligaments (ACL and MCL) are among the ligaments of the knee joint.

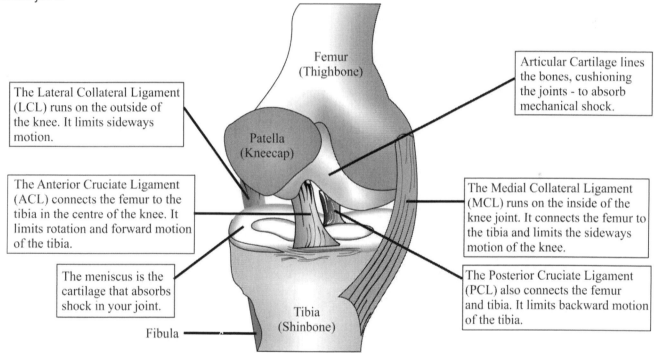

Fig. 9.1 – Structure of knee joint

Common injuries that occur in the knee joint are:

1. **Cruciate Ligament Damage – due to large forces during sporting activity.**

2. **Osteoarthritis – due to wearing-away of cartilage.**

1. Cruciate Ligament Damage

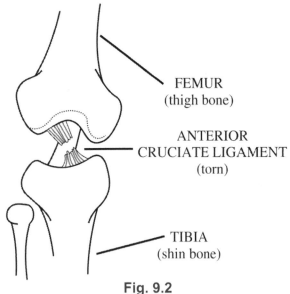

FEMUR
(thigh bone)

ANTERIOR
CRUCIATE LIGAMENT
(torn)

TIBIA
(shin bone)

Fig. 9.2

ACL injuries are common, especially if you play sports. The anterior cruciate ligament (ACL), is the most often injured ligament of the 4 ligaments in your knee. It is very painful and can be injured by hyper-extending the knee or twisting your knee inward.
This can be caused by:

- Sudden change of direction, causing a twist in your knee
- A rapid stop with a change in direction
- Slowing down while running
- Landing from a jump
- Contact or collision causing extreme knee hyperextension

When the ACL becomes injured you may hear and feel a "pop" in your knee. In most cases, this is the ligament tearing. If this does happen, be sure to apply the Rest, Ice, Compression, and elevation. Otherwise known as the RICE technique.

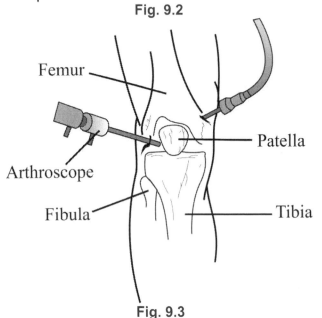

Femur

Patella

Arthroscope

Fibula

Tibia

Fig. 9.3

The patient is given anaesthesia, small punctures are made into the knee joint.

A small camera and small instruments on the end of long narrow tubes, introduced into the knee through small incisions (keyholes).

Remove all of the existing damaged ACL. This is done with a motorized device which is called a shaver.

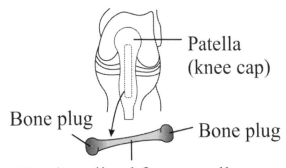

Patella
(knee cap)

Bone plug

Bone plug

Tendon sliced from patella
Fig. 9.4

Skin incisions (cuts) are made and the patellar tendon is identified. A slice of the tendon is harvested with a bone block at each end of the tendon. This slice of tendon is called the graft. The advantage of using the person's own tissue is that there will be **no tissue rejection**. The tissue is taken from around the knee joint, so that the procedure can be completed with a **single surgery.**

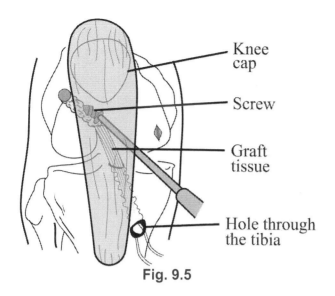

Knee cap

Screw

Graft tissue

Hole through the tibia

Fig. 9.5

Tunnels are drilled into the femur and tibia. The bone plugs of the graft will fit into these tunnels.

The graft is then placed through the tibia, through the knee joint, and into the femoral drill hole. Screws, staples or sutures are used to fix the bone plugs. The biological screws will dissolve in about two years.

Tendons are less elastic than ligaments. So physiotherapy will be needed to gently stretch the tendons and regain full movement. Overstretching must be avoided to prevent damage to the tendon.

Advantages of key hole surgery over open surgery
- A smaller incision is made, so there is less bleeding.
- There is a lower risk of infection.
- Less scarring occurs.
- The stay in hospital is shorter, as healing time is quicker. This saves money and reduces the chance of secondary infections.
- It is less likely to cause damage to cartilage, which can lead to osteoarthritis.

Disadvantages of key hole surgery over open surgery
- The procedure requires a high degree of training.
- The equipment is expensive, which could result in high surgical costs.
- It can only be used for certain types of surgery only.

2. Osteoarthritis – due to wearing-away of cartilage.
Over-training and sports injuries can result in greater wear and tear which may damage the cartilage in the knee joints and lead to osteoarthritis.

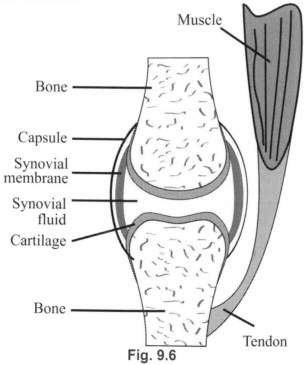

Muscle

Bone

Capsule

Synovial membrane

Synovial fluid

Cartilage

Bone

Tendon

Fig. 9.6

Osteoarthritis is a joint disease that affects the cartilage. In osteoarthritis, the surface layer of cartilage breaks down and wears away. This allows bones under the cartilage to rub together, causing pain, swelling and loss of motion of the joint. Over time, the joint may lose its normal shape.

Also, bone spurs – small growths called osteophytes – may grow on the edges of the joint. Bits of bone or cartilage can break off and float inside the joint space. This causes more pain and damage.

Osteoarthritis is common in knee joints because they are the body's primary weight-bearing joints.

Healthy cartilage covering the ends of the bones allows bones to glide over one another, and it absorbs energy from the shock of physical movement. Repetitive load-bearing can cause damage to the cartilage lining the joint. Sudden twisting and excessive force can cause injury to the cartilage, or this can take place over time with repeated use. As cartilage is not supplied with blood, it is very slow to repair itself when damaged. Once the cartilage is damaged, the protection it gives the joint decreases. This can lead to more rapid degeneration, resulting in osteoarthritis.

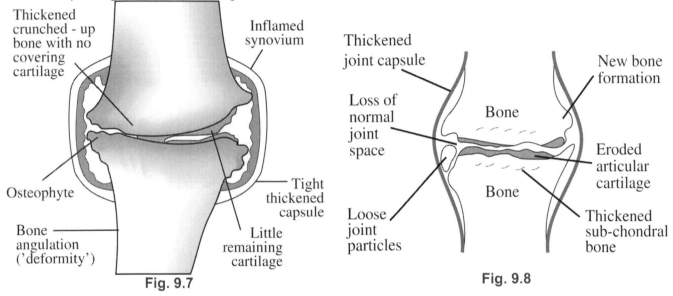

Fig. 9.7

Fig. 9.8

Keyhole surgery can be used to remove pieces of bone (bone debris) from the joint.

Prosthesis

A prosthesis is a device designed to replace a missing part of the body, or to make a part of the body work better. For many years, the most successful treatment for severe arthritis of the knee has been total knee replacement. This means removing all the surfaces of both compartments and replacing them with a metal implant on the lower end of the femur and a plastic surface on the tibia.

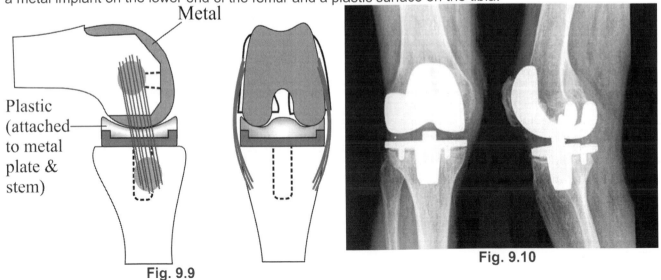

Fig. 9.9

Fig. 9.10

Ethical implications of using performance-enhancing substances by athletes

Why doping may be unethical?	Why doping may be considered ethical?
Athletes have a right of access to fair competition (doping is **unfair to those who do not do it**). Doping might be good (from the point of view of success in competition) for the few athletes that do it, but is bad for the many that do not.	Athletes have the right to achieve the best performance they can. Doping gives people a chance to be as good as their potential allows. It removes "unfair" genetic advantages.
Athletes have the right to be protected from harmful drugs. Sports governing bodies have a duty to ensure that athletes do not take drugs. It is unethical for athletes to play against the rules.	Athletes have a duty to sponsors to achieve their best performances. Economic benefits are too lucrative to ignore.
Many athletes who take drugs are not really doing so intentionally or willingly. They are usually under pressure from coaches to do so. Many athletes may not have given their informed consent.	It should be up to the individual athlete to decide for themselves whether they will take drugs to improve their performance or not. Especially if the drug is not banned and is not harmful.
Many spectators are disappointed to find that a successful athlete has taken drugs. Athletes know that there are rules against taking certain drugs. An honest and sincere athlete follows the rules.	Individuals have different 'codes of ethics'. If you believe the anti-doping rules are pointless or misguided, then there is no point in following the rules.
Many drugs have harmful side-effects. These drugs are often taken by athletes without medical supervision, which could lead to irreversible harm.	

The table below summarises the effect of a few performance enhancing drugs. It is not necessary to memorise the effects of these drugs.

Drug	Effect on physiology	Effect on performance	Side-effects
Erythropoietin (EPO)	EPO causes the bone marrow to generate extra red blood cells.	Extra blood cells mean the blood can carry extra oxygen. This increases the level of work the body can sustain through aerobic respiration (aerobic threshold).	Increased haemocrit increases blood viscosity. This causes strain on the heart and can lead to infarction
Creatine (Taken as a dietary supplement to increase the amount of creatine phosphate in the muscles).	Creatine combines with phosphate to form Creatine Phosphate (CP). CP can phosphorylate ADP, re-generating ATP.	Because ATP is re-generated without using the respiratory pathways, theoretically it should decrease recovery time. The use of creatine supplements combined with heavy weight training has been associated with increases in muscle mass and maximal strength.	Diarrhoea, vomiting, liver damage and kidney damage.
Testosterone	Binds to androgen receptors in target cells and increases transcription of anabolic proteins (growth proteins) such as actin & myosin.	Muscle mass increases, which makes the athlete more powerful. It also decreases recovery time.	Agression, decreased sex drive, infertility, skin problems, acne, shrunken testicles

Chapter Ten
Photoreception in Plants

Phytochrome is a photoreceptor pigment found in **seeds**, **stems** and **buds** of plants. It is **bluish-green** in colour. A phytochrome molecule consists of a protein component bonded to a non-protein light absorbing pigment molecule. Phytochrome exists in **two inter-convertible forms**: **Phytochrome red (P_r/P660)** and **Phytochrome far red (P_{fr}/P730)**.

Phytochromes help plants to respond to photoperiod, in the following ways:

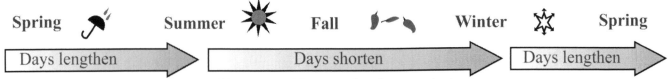

Fig. 10.1 – Changes in photoperiod

1. Phytochromes affect the germination of seeds.

P_{fr} stimulates germination. This ensures that seeds germinate only when long durations of light are available. This increases chances of survival of the seedling. Eg: only seeds just below the surface (which receive some light) will germinate, ensuring that they emerge from the soil and start to photosynthesise even before food stored in the cotyledons is used up.

● Seeds falling on forest floor, below the canopy, will not germinate until a large tree falls down. The seeds on forest floor act as a reserve and ensure that some plants survive even if there is a disastrous year in which adult plants cannot produce seeds.

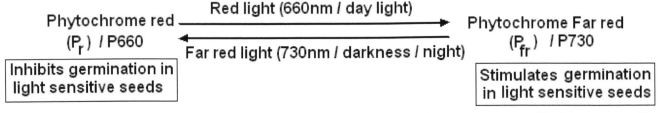

This mechanism is used by seeds to **start germinating in the spring or summer**. During summer the days are longer than in the winter. More P730 accumulates in the seeds and stimulates germination. On the other hand, **seeds do not germinate in winter** because there would be a higher concentration of P660 in the seeds. Remember that P660 inhibits germination. If seeds begin to germinate in winter, there is little chance of the seedling surviving the harsh environmental conditions of winter.

Note: the conversion into P_r is more rapid in far red light than in darkness.

2. Phytochromes influence the timing of flowering.

Phytochromes provide a mechanism to detect the length of daylight or night (photoperiod). This influences the timing of flowering in some plants.

a. **Long day plants (LDP)** flower only in summer, when the days are longer than nights. Flowering stimulated by P_{fr}. Examples: strawberries, oats, poppies and lettuce.

b. **Short day plants (SDP)** flower only in autumn, when the days are shorter than nights. Flowering stimulated by P_r. Examples: chrysanthemums and poinsettias.
c. **Day neutral plants (DNP)** are unaffected by the length of the day. **Example:** cucumbers, tomatoes and pea plants.

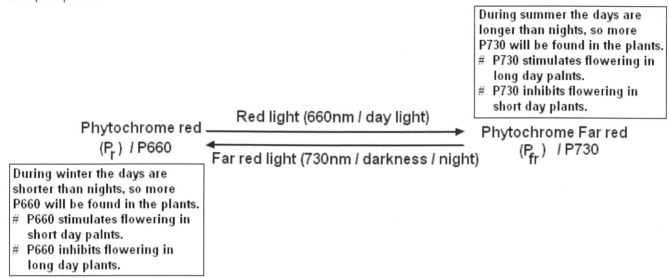

3. **Greening** refers to the changes that a young seedling undergoes as soon as it emerges from the soil.

The exposure to light stimulates the production of P_{fr}, which promotes the formation of chlorophyll, leaf development and unrolling of primary leaves. It also inhibits the elongation of internodes.
Etiolation is the opposite of greening. It occurs when plants are kept in complete darkness. P_r is produced in darkness and brings about the elongation of Internodes and yellowing of leaves. The elongation of stem helps the plant to grow rapidly upward in search of light.
Phytochromes and gene regulation *(refer to topic on gene regulation)*
Phytochromes may bind to transcription factors and **promote** or **inhibit** the formation of the transcription initiation complex.

For example: P_{fr} activates the proteins needed for the initiation of the light-regulated genes in young seedlings. This switches on the genes for the enzymes that produce chlorophyll. This promotes greening.

Other environmental cues:
Geotropism: response to gravity;
Thigmotropism: response to touch;
Phototropism: response to light;

CHAPTER ELEVEN
STRUCTURE AND FUNCTION OF NEURONES

Learning outcomes: by the end of this chapter you should be able to

Edexcel Syllabus Spec 3: Describe the structure and function of sensory, relay and motor neurones including the role of Schwann cells and myelination.

Edexcel Syllabus Spec 4: Describe how a nerve impulse (action potential) is conducted along an axon including changes in membrane permeability to sodium and potassium ions and the role of the nodes of Ranvier.

Structure of neurones

Neurones are nerve cells which carry electrochemical impulses from one part of the body to another. Every neurone consists of three main regions

Dendron: The Dendron receives stimulus or neurotransmitters from other cells. It carries impulses towards the cell body. The smaller branches which receive and transmit impulses to the dendron are called dendrites.

Cyton: The cyton or cell body contains most of the cell organelles and nucleus. It connects the dendron to axon.

Axon: The axon is an extension from the cell body. It carries impulses away from the cyton. The **axon endings** usually form **synapses** with the dendrites of other neurones.

Impulses always travel in one direction along a neurone – **from dendron** to **axon, across the cyton.**
The structure and function of the three basic types of neurones are shown and described below.

A. Sensory neurone: sensory neurones transmit **impulses from the sense organs to the central nervous system (brain or spinal cord).**

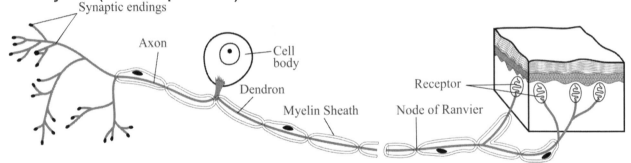

Fig. 11.1 – Sensory neurone attached to a receptor cell

B. Motor neurone: motor neurones always transmit impulses **from** the **central nervous system** (brain or spinal cord) **to** the **effector organs** (muscles or glands).
There are many dendrons but a single long axon which is myelinated. They are sometimes referred to as multi-polar neurones as they have many dendrites.

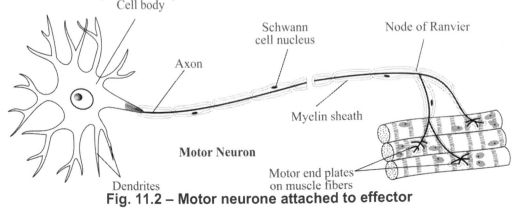

Fig. 11.2 – Motor neurone attached to effector

C. Relay neurone: these are also called connector neurones or intermediate neurones, they are non-myelinated. The main function of these neurones is to transmit impulses from one neurone to another. In the brain and spinal cord a single relay neurone connects with many different neurones and helps to analyse information received from many neurones. This is referred to as **summation.** The connections are made by **synapses.**

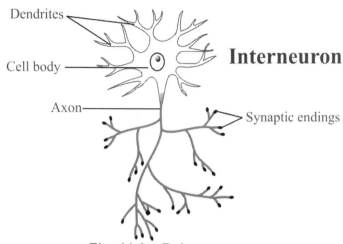

Fig. 11.3 – Relay neurone

Factors affecting speed of nerve impulse
  ✓ Diameter of the axon. (Speed is proportional to diameter)

  ✓ Myelination - In myelinated axons the action potential travels from one node of Ranvier to the next. This is called the SALTATORY EFFECT, which speeds up transmission of impulses.

In a sensory neurone, the dendron is longer than the axon and is insulated by a **myelin sheath**, formed of **Schwann cells.** The Schwann cells are wrapped around the axon to form many layers of insulating material composed of 70% lipid and 30% proteins. Between adjacent Schwann cells there are gaps called the **Nodes of Ranvier.** These regions are not insulated and can transmit nerve impulses. The nodes of Ranvier speed up the transmission of nerve impulses, as the impulse jumps from one node of Ranvier to the next. This is called the **Saltatory effect.**

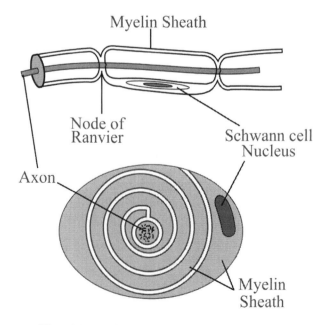

Fig. 11.4 – Structure of myelin sheath

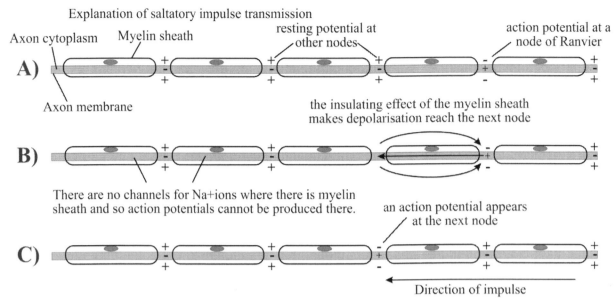

Explanation of saltatory impulse transmission

Fig. 11.5 – Saltatory conduction in a myelinated neurone

A – The first node of Ranvier is depolarised (action potential);

B – A local circuit forms between the first node of Ranvier and the adjacent Node of Ranvier;

C – The action potential has now formed at the second node (from the right) of Ranvier and the action potential moves along the neurone from right to left. The myelin sheath is insulating and allows action potentials to form only at the nodes of Ranvier.

Components of the neurone cell membrane.

- **Sodium potassium pumps:** These are carrier proteins which pump 3Na$^+$ ions out and 2K$^+$ ions into the cell, by active transport.
- **Sodium channels** are permeable to sodium ions. These are also referred to as voltage gated channels as they open and close based upon the potential in the cell. These gates close at +40 mV.
- **Potassium channels** are permeable to potassium ions. These are also referred to as voltage gated channels as they open and close based upon the potential in the cell. These gates open at +40 mV and close at -80 mV.

The cytoplasm contains negatively charged proteins. These ions remain in the cytoplasm. Changes in the potential of the membrane are brought about by movement of sodium and potassium ions across the membrane.

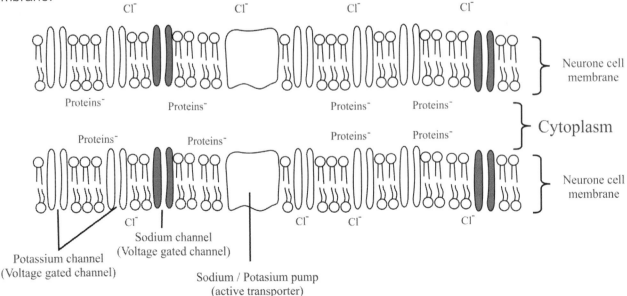

Fig. 11.6 – Components of neurone membrane and cytoplasm

Resting potential (-70 mV): The Na-K pump, pumps $3Na^+$ ions out and $2K^+$ ions into the neurone by **active transport**. The membrane is more permeable to K^+ ions than to Na^+ ions (as there are more potassium ion channels than sodium ion channels per unit area of the membrane). So, K^+ ions diffuse out of the neurone more rapidly than Na^+ ions diffuse in. This results in the **net removal of positive charges** from the cell. This causes the inside of the cell to become more negative. When the membrane reaches a potential of -70mV the movement of the K^+ reaches equilibrium as the opposing pulls of the diffusion gradient and the electrical gradient balance out at around –70 mV.
This is called the resting potential. The membrane is said to be **polarised**.

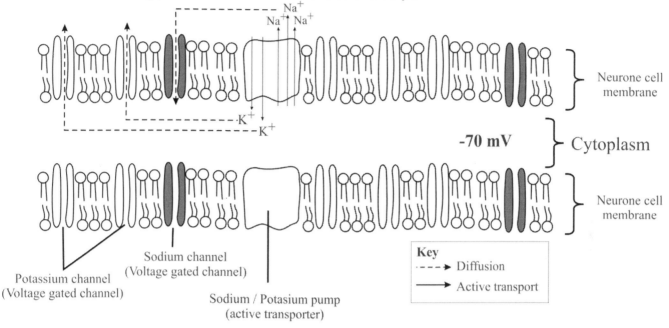

Fig. 11.7 – Maintenance of a resting potential

Most of the Na^+ ions are now outside the cell and most of the K^+ ions are inside the cell. The sodium and potassium gates are **closed** and only **slightly permeable**.

Action potential (+40 mv): The action potential is generated by the opening of the sodium channels. Na^+ ions enter into the cytoplasm down the electrochemical and concentration gradient, until the potential reaches +40 mV. The potential of +40 mV causes the Na^+ gates to close and the influx of Na^+ ions stops. This is called as the action potential and the membrane is said to be **depolarised**.

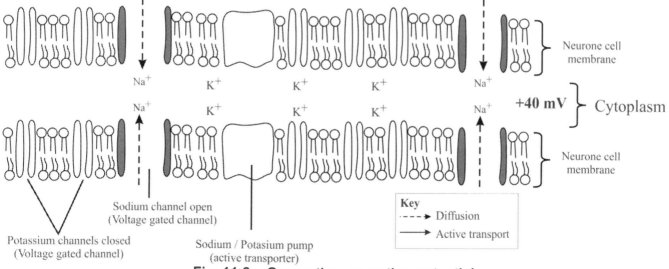

Fig. 11.8 – Generating an action potential

Note: Most of the sodium ions and potassium ions are inside the cell.

Hyperpolarisation and repolarisation

Once the action potential is reached, the sodium gates close and the potassium gates open. Na^+ ions cannot move into the cell but K^+ ions move out rapidly causing repolarisation. There is an overshoot of K^+ ions out of the cell, which causes the potential to fall lower than -70 mV. This is called hyperpolarisation (- 80mV). The potassium gates close at this potential. Hyperpolarisation ensures that the impulse passes along the neurone in one direction only. It is also responsible for the refractory period.

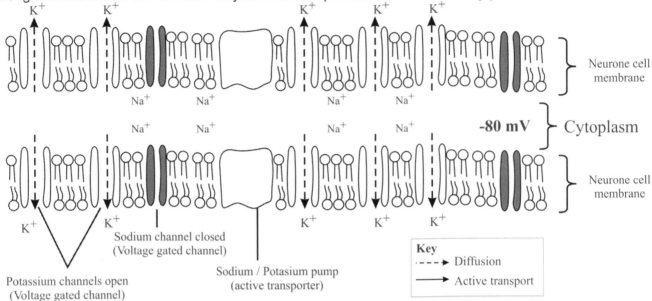

Fig. 11.9 – Hyperpolarisation of the Membrane

Closure of the K^+ ion gates, the action of the sodium-potassium pump and diffusion of potassium ions restores the resting potential.

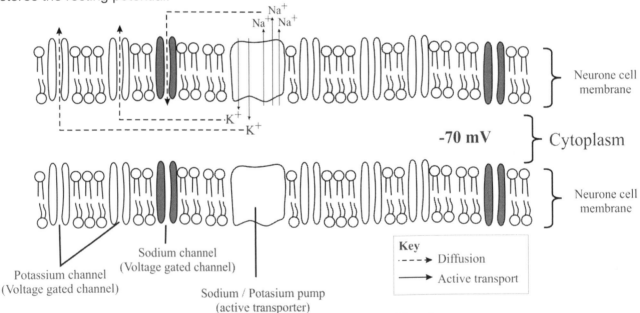

Fig. 11.10 – Repolarisation of the Membrane

Threshold potential: It is the minimum depolarisation that is needed to generate an action potential. If the membrane depolarises up to the threshold potential, then more Na^+ channels open (This is an example of positive feedback) and an action potential is generated. If the depolarisation does not reach the threshold potential, then no action potential is generated. This is often referred to as the **all or nothing principle** of nerve impulse transmission.

Refractory period: This is the time period required for the membrane to regain the resting potential, after firing an action potential. The membrane is hyperpolarising and the K^+ ions are moving out of the cell. This ensures that the impulse can flow along the neurone in one direction only.

The graph in fig 11.11 shows changes in the membrane potential during a nerve impulse.

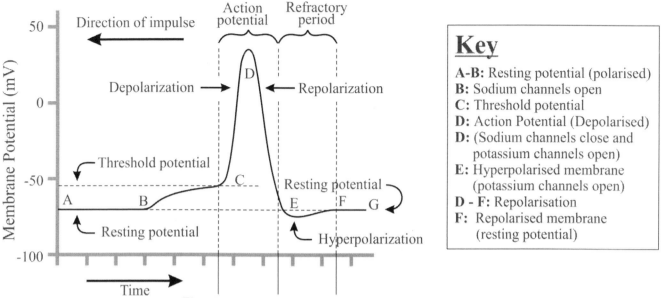

Fig. 11.11 – Changes in membrane potential

A nerve impulse is the propagation (movement) of an action potential (a wave of depolarization) along a neurone.

Key
A-B: Resting potential (polarised)
B: Sodium channels open
C: Threshold potential
D: Action Potential (Depolarised)
D: (Sodium channels close and potassium channels open)
E: Hyperpolarised membrane (potassium channels open)
D - F: Repolarisation
F: Repolarised membrane (resting potential)

Even though resting potential, action potential, hyperpolarisation and repolarisation are described separately it must be kept in mind that all these processes occur in close succession in a neurone. The **entire neurone does not become depolarised or repolarised.** Instead the impulse moves as a wave of depolarisation along the neurone. Refer to fig. 11.12 to understand how a nerve impulse travels along a neurone. **Local currents drive the action potential from one point of the neurone to the next.** The **local currents** are responsible for the **opening of the sodium channels**. The direction of transmission of an impulse is always from a point of action potential towards the resting potential, away from a point of hyperpolarisation.

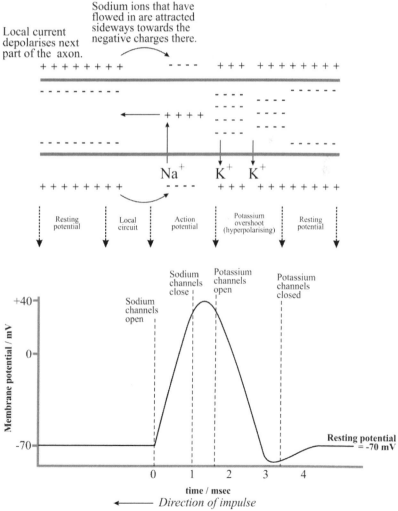

Fig. 11.12 – Formation of local circuits

CHAPTER TWELVE
SYNAPTIC TRANSMISSION

Learning outcomes: by the end of this chapter you should be able to
Edexcel Syllabus Spec 5: *Describe the structure and function of synapses, including the role of neurotransmitters, such as acetylcholine.*

Synapse is a gap between the end of one neurone and the beginning of another neurone. If the gap is less than 2nm then electrical impulses are transmitted from one neurone to the next. But, if the gap (cleft) is larger (up to 20nm) then chemical (neurotransmitters) transmission operates.

- Synapses consist of:
  - **presynaptic ending** (where neurotransmitters are made)
  - **post synaptic ending** (has neuroreceptors in the membrane)
  - **synaptic cleft**
- Action potentials **cannot** cross the synaptic cleft
- Nerve impulse is carried by **neurotransmitters**

The sequence of events involved in the transmission of an impulse across a synapse is illustrated in the diagrams below. This synapse involves acetylcholine as a transmitter substance. Hence it is called a **cholinergic synapse**.

a. Arrival of action potential at the presynaptic knob makes it more permeable to calcium ions, which diffuse into the presynaptic knob from the synaptic cleft.

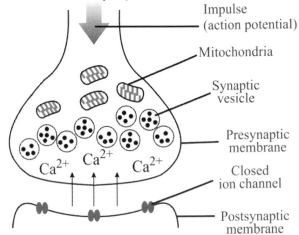

Fig.12.1 – Action potential arrives at the presynaptic membrane

b. This causes movement of synaptic vesicles towards the presynaptic membrane. These vesicles fuse with the presynaptic membrane and release neurotransmitter substances into synaptic cleft by exocytosis.

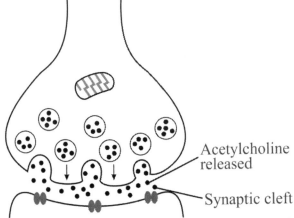

Fig.12.2 – Neurotransmitter (acetylcholine) released into the synaptic cleft

c. Neurotransmitter (acetylcholine) molecules diffuse across the synaptic cleft and bind with receptors on the postsynaptic membrane. The binding of neurotransmitter (acetylcholine) with the receptors on the postsynaptic membrane cause the membrane to become permeable to sodium ions. If the membrane reaches the threshold potential, then many sodium channels open and there will be rapid influx of sodium ions into the post synaptic membrane. The rapid influx of sodium ions causes the generation of an action potential in the postsynaptic neurone.

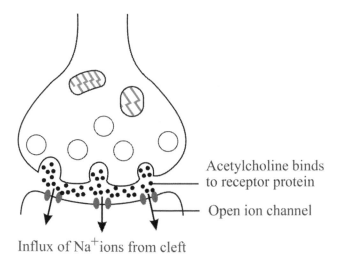

Acetylcholine binds to receptor protein

Open ion channel

Influx of Na$^+$ ions from cleft

Fig.12.3 – Depolarisation of post-synaptic membrane

d. The neurotransmitter substance is then removed from the receptors on the post synaptic membrane by enzymes. The enzyme cholinesterase hydrolyses acetylcholine to choline and ethanoic acid, which are inactive as neurotransmitters. This enables the postsynaptic membrane to repolarise, so that the next action potential can be generated. Choline and ethanoic acid are reabsorbed into the presynaptic neurone and are used to resynthesise acetylcholine. Mitochondria are used to release energy for the synthesis of neurotransmitters.

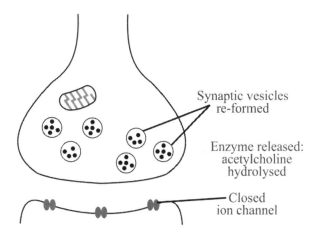

Synaptic vesicles re-formed

Enzyme released: acetylcholine hydrolysed

Closed ion channel

Fig.12.4 – repolarisation of post-synaptic membrane

Summary of synaptic transmission

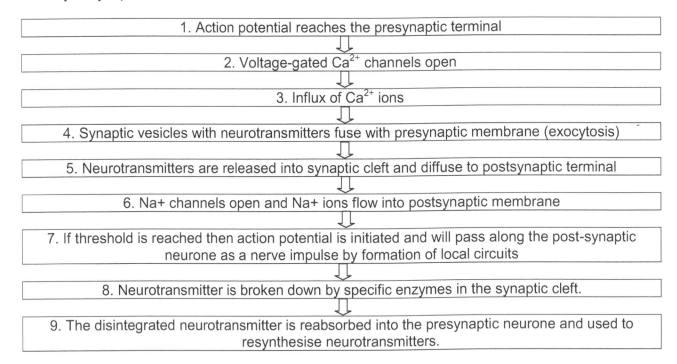

1. Action potential reaches the presynaptic terminal

⇓

2. Voltage-gated Ca^{2+} channels open

⇓

3. Influx of Ca^{2+} ions

⇓

4. Synaptic vesicles with neurotransmitters fuse with presynaptic membrane (exocytosis)

⇓

5. Neurotransmitters are released into synaptic cleft and diffuse to postsynaptic terminal

⇓

6. Na+ channels open and Na+ ions flow into postsynaptic membrane

⇓

7. If threshold is reached then action potential is initiated and will pass along the post-synaptic neurone as a nerve impulse by formation of local circuits

⇓

8. Neurotransmitter is broken down by specific enzymes in the synaptic cleft.

⇓

9. The disintegrated neurotransmitter is reabsorbed into the presynaptic neurone and used to resynthesise neurotransmitters.

Functions of synapses

1. Synapses make sure that the **flow of impulses is in one direction only.** This is because the vesicles containing the transmitter are only in the presynaptic membrane and the receptor molecules are only on the postsynaptic membrane.
2. They allow **integration:** impulses can be sent to or received from several neurones. An impulse travelling down a neurone may reach a synapse which has several post synaptic neurones, all going to different locations. The impulse can thus be dispersed. This can also work in reverse, where several impulses can converge at a synapse.
3. They allow '**summation**' to occur. Synapses require the release of sufficient transmitter into the cleft in order for enough of the transmitter to bind to the postsynaptic receptors and the impulse to be generated in the postsynaptic neurone.

In **spatial summation**, several presynaptic neurones converge at a synapse with a single post synaptic neurone. The release of neurotransmitter substance from the presynaptic neurone by a single impulse is not always enough to generate an action potential in the postsynaptic neurone. Normally several vesicles have to be released before there is enough transmitter substance in the synaptic cleft to propagate an action potential. This is possible by multiple impulses. This addition effect is called summation.

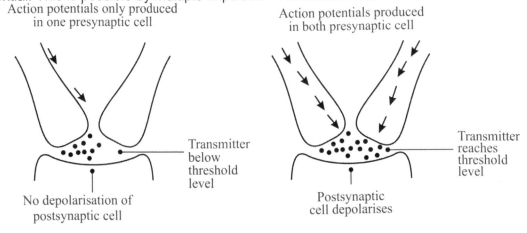

Fig.12.5 – Spatial summation

In **temporal summation** there is only one presynaptic and one postsynaptic neurone but the frequency of impulses reaching the synapse is important. Each action potential that arrives at the pre-synaptic membrane will cause a number of vesicles to release their transmitter. A number of action potentials are required before there is enough transmitter to initiate an action potential in the post-synaptic cell. This is called temporal summation and is seen to occur in the cone cells of the retina.

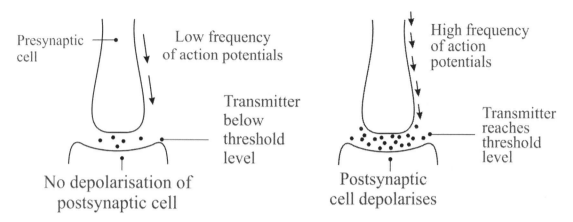

Fig.12.6 – Temporal summation

A number of pre-synaptic neurones may form synapses with one post-synaptic neurone. Action potentials arriving in each pre-synaptic neurone will release transmitter, which build up to the threshold level and triggers a postsynaptic impulse. This is spatial summation. It is responsible for synaptic convergence and is seen to occur in rod cells of the retina.

4. Synapses allow the 'filtering out' of continual unnecessary or unimportant background stimuli. If a neurone is constantly stimulated (e.g. clothes touching the skin) the synapse will not be able to renew its supply of transmitter fast enough to continue passing the impulse across the cleft.   This 'fatigue' places an upper limit on the frequency of depolarisation.

Excitatory and inhibitory synapses.

Excitatory synapses cause depolarization of the postsynaptic membrane. The examples above are **excitatory postsynaptic potentials**.

Inhibitory synapses: the neurotransmitters released from the presynaptic neurone will not cause depolarization of the postsynaptic neurone. Instead they cause the postsynaptic membrane to become more negative than usual (hyperpolarized / -90mV), by opening of chloride ion channels. This makes the neurone less likely to trigger an action potential. This is called the **inhibitory postsynaptic potential.**
Example: glutamate is a neurotransmitter released from the rod cells. Glutamate inhibits the depolarisation of the bipolar neurones. So, glutamate in the retina acts as an inhibitory neurotransmitter.

Note: The transmitter substances are not inherently excitatory or inhibitory. For example, acetylcholine has an excitatory effect at most neuromuscular junctions and synapses, but has an inhibitory effect on neuromuscular junctions in cardiac muscle and gut muscle. These opposing effects are determined by events occurring at the post synaptic membrane. The molecular properties of the receptor sites determine which ions enter the postsynaptic cell, which in turn determines the nature of the change in postsynaptic potentials.

Learning outcomes: by the end of this chapter you should be able to

Edexcel Syllabus Spec 6: *Describe how the nervous systems of organisms can detect stimuli with reference to rods in the retina of mammals, the roles of rhodopsin, opsin, retinal, sodium ions, cation channels and hyperpolarisation of rod cells in forming action potentials in the optic neurones.*

Edexcel Syllabus Spec 7: *Explain how the nervous systems of organisms can cause effectors to respond as exemplified by pupil dilation and contraction.*

Structure of retina

The eyes are the most important sense organs. We receive about 80% of information from the surroundings through the eyes. Photosensitive pigments in the eye convert light energy into electrical signals, in the form of nerve impulses. The process by which a stimulus is converted into a nerve impulse is called **Transduction**. These nerve impulses are then sent to the **visual cortex** of the brain for interpretation. An image is then perceived.

There are two types of photosensitive pigments found in the retina of the eye.

Rhodopsin: this is the photosensitive pigment found in the **rod cells** of the retina. It is sensitive to light of **low intensity / dim light**.

Iodopsin: this is the photosensitive pigment found in the **cone cells** of the retina. It is sensitive to the **wavelength of light / colour vision**.

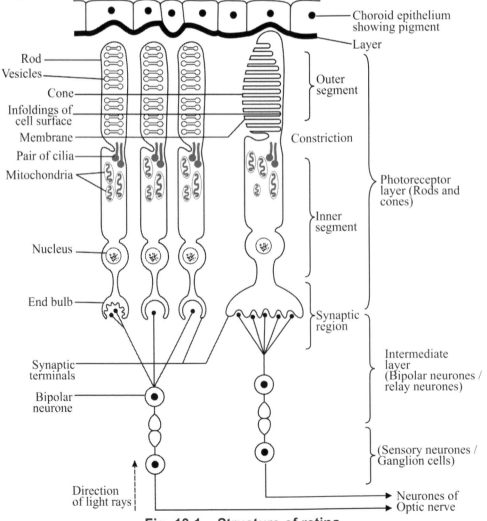

Fig. 13.1 – Structure of retina

Fig. 13.1 shows the arrangement of rods and cones in the retina of the eye. Notice that there are three layers of cells – the outermost layer of rods and cones (towards the choroid); the layer with bipolar neurones and the innermost layer with ganglion cells (towards the vitreous humour).

Structure of a rod cell

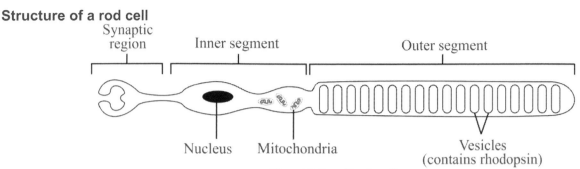

Fig. 13.2 – Rod cell

The detection of light is carried out on the membrane disks (vesicles) in the outer segment. These disks contain thousands of molecules of **rhodopsin**, the photoreceptor molecule. Rhodopsin consists of a membrane-bound protein called <u>opsin</u> and a covalently-bound prosthetic group called **retinal**. Retinal is derived from vitamin A, and a dietary deficiency in this vitamin causes night-blindness (poor vision in dim light). Retinal is the light-sensitive part, and it can exists in 2 forms: a *cis*-form and a *trans*-form. **Light causes *cis*-retinal to be converted into *trans*-retinal, which splits away from opsin. *Trans*-retinal is reconverted into *cis*-retinal by an enzyme retinal isomerase. The cis-retinal recombines with opsin to form Rhodopsin. ATP is used for this process.** The process is illustrated in fig. 13.3.

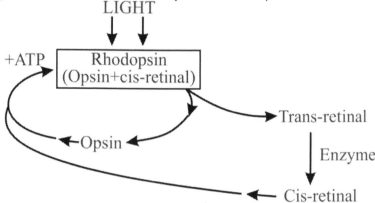

Fig. 13.3 – Conversion cycle of retinal

In the dark retinal is in the *cis* form, but when it absorbs a photon of light it quickly switches to the *trans* form. This causes the *trans*-retinal to split from the opsin. This process is called **bleaching**. The reverse reaction (*trans* to *cis* retinal) requires an enzyme reaction and is very slow, taking a few minutes. This explains why you are initially blind when you walk from sunlight to a dark room. Some time is needed to resynthesise rhodopsin. This is called as the **dark adaptation**.

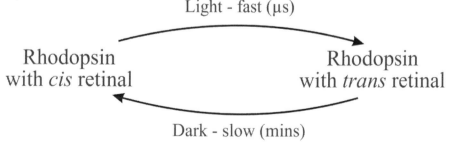

Fig. 13.4 – Conversion rate compared

Rod cell membranes contain a special sodium channel that is controlled by rhodopsin. Rhodopsin with *cis* retinal opens it and rhodopsin with *trans* retinal closes it. The synapse with the bipolar cell is an **inhibitory synapse**, so the neurotransmitter (glutamate) **stops** the bipolar cell making a nerve impulse.

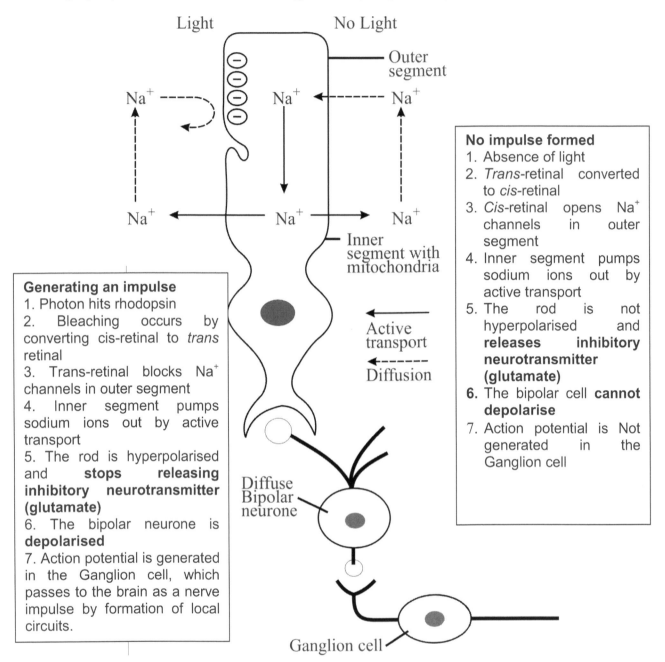

Light **No Light**

Generating an impulse
1. Photon hits rhodopsin
2. Bleaching occurs by converting cis-retinal to *trans* retinal
3. Trans-retinal blocks Na⁺ channels in outer segment
4. Inner segment pumps sodium ions out by active transport
5. The rod is hyperpolarised and **stops releasing inhibitory neurotransmitter (glutamate)**
6. The bipolar neurone is **depolarised**
7. Action potential is generated in the Ganglion cell, which passes to the brain as a nerve impulse by formation of local circuits.

No impulse formed
1. Absence of light
2. *Trans*-retinal converted to *cis*-retinal
3. *Cis*-retinal opens Na⁺ channels in outer segment
4. Inner segment pumps sodium ions out by active transport
5. The rod is not hyperpolarised and **releases inhibitory neurotransmitter (glutamate)**
6. The bipolar cell **cannot depolarise**
7. Action potential is Not generated in the Ganglion cell

Fig. 13.5 – Functioning of rod cells

Cones work in exactly the same way, except that they contain the pigment **Iodopsin**, which is found in 3 different forms; red-sensitive, blue-sensitive and green-sensitive. This gives us colour vision.

Synaptic convergence and its effects
There are about 1.2×10^8 rod cells in the retina. The rod cells are evenly distributed in the retina except at the fovea and the blind spot, where there are no rod cells. The rod cells enable vision low light intensities and are not sensitive to colours (wavelength of light). Rod cells are not tightly packed in the retina and **several rod cells synapse with a single bipolar neurone**. This is called **synaptic convergence** (about 150 rod cells synapse with a single diffuse bipolar neurone). The effect of synaptic

convergence is that the rod cells do not give a particularly clear picture (**low visual acuity**). However, synaptic convergence causes the rod cells to be very sensitive to low light intensity and to movements in the field of vision, because several small stimuli in the rod cells can stimulate an action potential in the ganglion cells.

Visual acuity of cone cells

Cone cells are tightly packed in the fovea. There are about 6×10^6 cones. Each cone cell is connected to an individual bipolar neurone (temporal summation). This means that more information is being sent to the brain. This provides a picture of great visual acuity. Since the cones are highly concentrated on the fovea, it is necessary for the image to be focused on the fovea to give a clear image of high visual acuity.

Pupil dilation and contraction

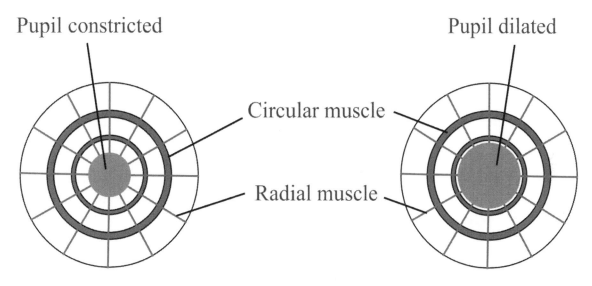

High light intensity
Circular muscles: contracted
Radial muscles: relaxed
Pupil diameter: small

Low light intensity
Circular muscles: relaxed
Radial muscles: contracted
Pupil diameter: large

Fig. 13.6 – Pupil reflex

The pupil dilation and constriction is regulated by a part of the nervous system called the autonomic nervous system (ANS). The ANS is part of the peripheral nervous system and it controls many organs and muscles within the body. In most situations, we are unaware of the workings of the ANS because it functions in an involuntary, reflexive manner.

To operate well in different light conditions the eye must be able to control the amount of light entering the eye. It does this by changing the diameter of the pupil. The changes in the size of the pupil are brought about by a reflex arc so it involves no conscious thought. The light <u>receptors</u> in the retina detect the intensity of the light and send this information to the brain (thalamus) via the optic nerve (Sensory Neuron). Then impulses from the brain are transmitted to the iris muscles (effectors) by the oculomotor nerve fibres (motor neurones).

Component	Structure
Stimulus	Light intensity
Receptor	Rods and cones in retina
Sensory neurones	Optic nerve
Coordinator	Brain (midbrain)
Motor nerves	Oculomotor nerve
Effector	Circular and radial muscles of iris
Response	Dilation or constriction of pupil

CHAPTER FOURTEEN
COMPARISON OF COORDINATION
IN PLANTS AND IN ANIMALS

Learning outcomes: by the end of this chapter you should be able to
Edexcel Syllabus Spec 8: *Compare mechanisms of coordination in plants and animals, ie nervous and hormonal, including the role of IAA in phototropism (details of individual mammalian hormones are not required).*

Growth in plants is coordinated by plant growth substances (PGS). These are produced in certain areas of the plant and transported to other parts where they can affect cell division, cell elongation and cell differentiation.

Auxins (Indole Acetic Acid – IAA) promote cell elongation. Auxins in coleoptile tips cause the coleoptiles to bend towards light (positive phototropism). The auxins move away from the illuminated side of the coleoptile and accumulate on the darker side, where it stimulates cell elongation (growth). The auxins soften the cell wall. The cell becomes less turgid and takes up more water, resulting in expansion of the cell. Due to the orientation (pattern of arrangement) of cellulose microfibrils in the cell wall, cell elongation occurs in the longitudinal direction.

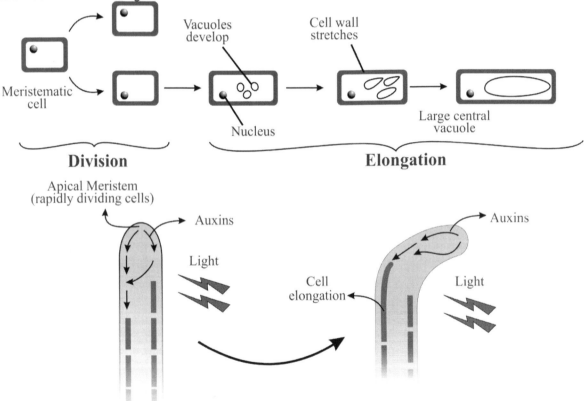

Fig. 14.1 – Effect of IAA on coleoptiles tips

Nervous control	Hormonal control
Transmitted by electrochemical impulses	Transmitted by chemical messengers (hormones)
Speed of transmission is very quick	Speed of transmission much slower
The effect is usually very short lived	The effect is usually long lasting

CHAPTER FIFTEEN
THE HUMAN BRAIN

Learning outcomes: by the end of this chapter you should be able to

Edexcel Syllabus Spec 9: *Locate and state the functions of the regions of the human brain's cerebral hemispheres (ability to see, think, learn and feel emotions), hypothalamus (thermoregulate), cerebellum (coordinate movement) and medulla oblongata (control the heartbeat).*

Structure of the brain

The brain and spinal cord are made up of neurones and glial cells (neuroglia). These neurones are interconnected by synapses to for an intricate network. The neuroglia provides physiological and physical support to the neurones.

The cerebrum is the largest part of the brain and consists of about 14 billion neurones. The outer part of the cerebrum consists of grey matter (cerebral cortex) and the internal part consists of white matter. Different regions of the cerebrum control different functions. The cerebrum consists of two hemispheres. The right cerebral hemisphere controls artistic abilities, creative thinking, music and recognition of shapes. The left cerebral hemisphere controls mathematical abilities, logical thinking, speech and writing.

The cerebellum is located posterior and below the cerebrum. It is smaller than the cerebrum and has a tree like, branched appearance.

The medulla oblongata connects the brain to the spinal cord.

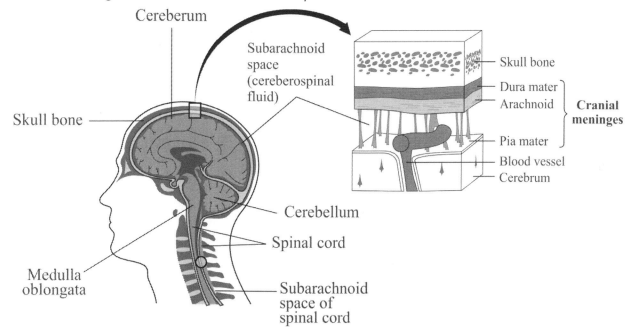

Fig. 15.1 – Structure of brain and spinal cord

The spinal cord is about 43 cm long and about the thickness of a pencil. The spinal cord is located in the neural canal of the vertebral column. The spinal cord is also made of neurones. But the arrangement of neurones is different from that of the brain - the spinal cord has white matter on the outside and grey matter on the inside.

Both the brain and spinal cord are surrounded by membranes called the meninges, in which cerebrospinal fluid circulates. The diagram above shows the structure of the meninges, brain and spinal cord. The cerebrospinal fluid serves as a medium of transport between the blood and neurones. It provides nourishment to the cells of the brain and spinal cord and to carry away wastes. It also provides lubrication.

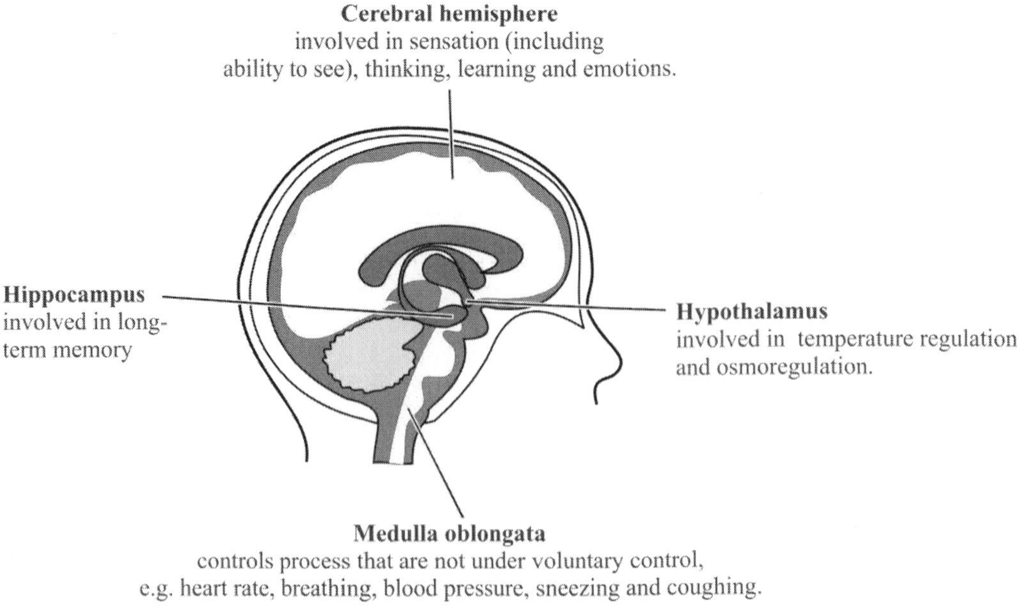

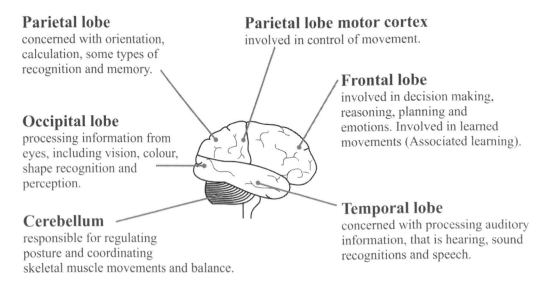

Cerebral hemisphere
involved in sensation (including
ability to see), thinking, learning and emotions.

Hippocampus
involved in long-
term memory

Hypothalamus
involved in temperature regulation
and osmoregulation.

Medulla oblongata
controls process that are not under voluntary control,
e.g. heart rate, breathing, blood pressure, sneezing and coughing.

Fig. 15.2 – Structure and function of brain

✓ **Cerebrum**: receives impulses, controls contraction of voluntary muscles, controls mental activities like speech, memory, emotions and other conscious activities.

✓ **Cerebellum**: maintenance of balance, posture of body and coordination of skeletal muscles.

✓ **Medulla oblongata**: controls rate of heartbeat, breathing rate, reflexes like coughing, sneezing and vomiting.

✓ **Hypothalamus**: plays an essential role in osmoregulation, maintaining body temperature and synthesis of hormones of posterior pituitary gland.

Parietal lobe
concerned with orientation,
calculation, some types of
recognition and memory.

Parietal lobe motor cortex
involved in control of movement.

Frontal lobe
involved in decision making,
reasoning, planning and
emotions. Involved in learned
movements (Associated learning).

Occipital lobe
processing information from
eyes, including vision, colour,
shape recognition and
perception.

Cerebellum
responsible for regulating
posture and coordinating
skeletal muscle movements and balance.

Temporal lobe
concerned with processing auditory
information, that is hearing, sound
recognitions and speech.

Fig. 15.3 – Structure and function of cerebellum and cerebrum

CHAPTER SIXTEEN
INVESTIGATING BRAIN STRUCTURE AND FUNCTION

Learning outcomes: by the end of this chapter you should be able to
Edexcel Syllabus Spec 10: *Describe the use of magnetic resonance imaging (MRI), functional magnetic resonance imaging (fMRI) and computed tomography (CT) scans in medical diagnosis and investigating brain structure and function.*

In the past, most of our information about the functions of the human brain was obtained from situations where parts of the brain had been damaged or missing at birth or as the result of illness or injury.

Some classic examples are discussed below:

1. Phineas Gage was a likeable, reliable, hardworking and very responsible American. An iron bar passed through his head during a dynamite explosion. It didn't kill him, but it destroyed much of the front part of the left-hand side of his brain. Gage could still walk, talk and carry on normal life activities, but his personality changed dramatically. He became impatient, irresponsible and unpleasant. He lost his job, and died 12 years later. Researchers at Harvard have used computer graphics along with images of the injury and the skull to show that the bar destroyed much of the connection between the left-side frontal lobes and the midbrain. Because of this Phineas Gage lost the ability to control his emotional behaviour.

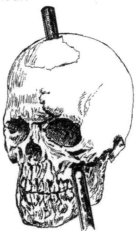

Fig. 16.1 – Phineas Gage' brain damage

2. In one particular disorder a man mistook his wife for a hat! This man had a disease affecting the visual areas of the brain. He could see and describe things but he had lost the ability to make the normal connections between what he saw and what the object was.

3. Another patient had a massive stroke which affected the deeper and back portions of her cerebral hemispheres. Her personality and ability to talk were unimpaired, but she completely lost the concept of the left-hand side, both of her own body and the world around her. She would eat only the right-hand portion of her food and applied make-up only to the right-hand side of her face. This shows that the location of function and awareness lies in different sides of the brain.

Brain imaging techniques provide a **non-invasive technique** for studying the functions of different regions of the brain. It also provides a valuable tool to investigate the development of the brain and critical windows. It has also made it possible to understand the progress of many brain diseases and how they affect the people who have them.

Technique	How it works	What it allows us to see
Surgery (Invasive technique) or deep brain stimulation	During brain surgery a local anaesthetic is often used. The skull is opened up or penetrated with a laproscopic probe. This allows the surgeon to ask the patient questions as he operates on their brain.	The patient can tell the doctor what he/she is feeling as the doctor stimulates parts of his/her brain. This can tell us a lot about the function of the brain. Deep brain stimulation is used for treatment of Parkinson's disease and 'suicide' headaches.
C T Scan (Computerised Tomography) or CAT scan (Computerised Axial Tomography) CT scan showing a stroke in the left cerebral hemisphere	Thousands of narrow-beam X-rays pass through the patient's head from a rotating source. The X-rays are collected on the other side of the head and their strength measured. The density of the tissue the X - ray passes through decreases the strength of the signal, and therefore, lets us work out what type of tissue is in the brain.	CT Scans show brain structures, **not** brain activity. CT scans only give "frozen" still images. However, they are very useful for picking up diseases, such as cancer, stroke and oedema. The resolution of the scan is low which means that very small brain structures can't be distinguished. Another disadvantage is that they use potentially harmful X-rays.
MRI Scan (Magnetic Resonance Imaging) 3D MRI scan of the head	Magnetic fields and radio waves are used to align protons in water molecules in the patient's brain. When the fields are switched off, the protons give out a little energy, which can be detected.	By recording the energy given out by protons we can build up a sequence of thin pictures of the types of tissues inside the brain. This can be fed into a computer, which uses the picture to build up a 3D image of the inside of the head. MRI Scans also only show brain structures, **not** brain activity. They also only give "frozen" still images. However, they are very useful for picking up diseases, such as cancer, stroke, oedema and brain damage.
fMRI Scan (functional Magnetic Resonance Imaging)	fMRI shows which parts of the brain are active during a particular task. Deoxygenated haemoglobin is more magnetic than oxygenated haemoglobin, which is virtually nonmagnetic. Oxyhaemoglobin does not absorb radio signals but deoxyhaemoglobin absorbs radio signals. The difference in oxyhaemoglobin content of the brain is used to rapidly produce a three dimensional (3D) image of the active and non active regions of the brain.	Similar to MRI, **but** the doctor not only knows what the tissues look like, but whether they are **active**. This is the only technique, that shows brain activity. It is therefore very useful in the study of brain function.

CHAPTER SEVENTEEN
CRITICAL WINDOW IN
DEVELOPMENT OF VISION

Learning outcomes: by the end of this chapter you should be able to
Edexcel Syllabus Spec 11: Discuss whether there exists a critical 'window' within which humans must be exposed to particular stimuli if they are to develop their visual capacities to the full.
Edexcel Syllabus Spec 12: Describe the role animal models have played in developing explanations of human brain development and function, including Hubel and Wiesel's experiments with monkeys and kittens.

Brain development

At birth, the human brain is still preparing for full operation. The brain's neurons exist mostly apart from one another. The brain's task for the first 3 years is to establish and reinforce connections with other neurons. These connections are formed when impulses are sent and received between neurons. These connections form synapses. As a child develops, the synapses become more complex, like a tree with more branches. During the first 3 years of life, the number of neurons stays the same and the number of synapses increases. After age 3, the creation of synapses slows until about age 10. Between birth and age 3, the brain creates more synapses than it needs. The **synapses that are used very frequently, become a permanent part of the brain**. The **synapses that are not used frequently are eliminated**. **"Neurones that fire together, wire together". "use it or lose it".** This is where experience plays an important role in wiring a young child's brain.

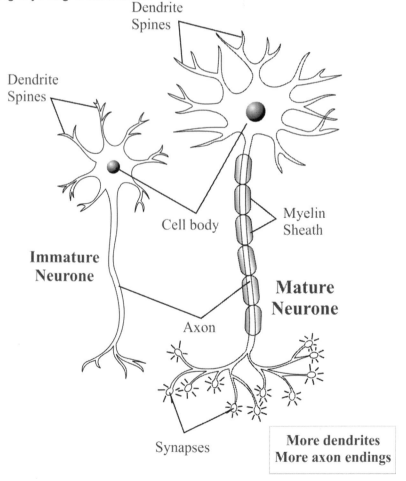

Fig. 17.1 – Changes in neurone structure during development

Research has shown that the brain development in children is influenced by the environment. There are definite periods of time when a child must have sensory or motor input if 'normal' development is to take place. These periods are called 'critical windows' or 'critical periods' or 'sensitive periods'.

During a critical window for visual development, there is a change of the structure, function, and organization of neurons in the visual cortex of the brain in response to visual stimulus from the environment. **This re-wiring of neurones is called as neural plasticity** (a change of the structure, function, and organization of neurons).

Neurones travel from the retina to the thalamus, through the optic nerve. These neurones synapse with other neurones in the thalamus. The thalamus sends impulses from the same area of the retina in the left and right eye to adjacent columns of cells in the visual cortex. This helps to create a map of the retina on the visual cortex.

Since humans have binocular vision (both eyes focus on the same object at the same time), the adjacent columns receive the same stimulus and are of equal width. This helps in full visual development of the individual.

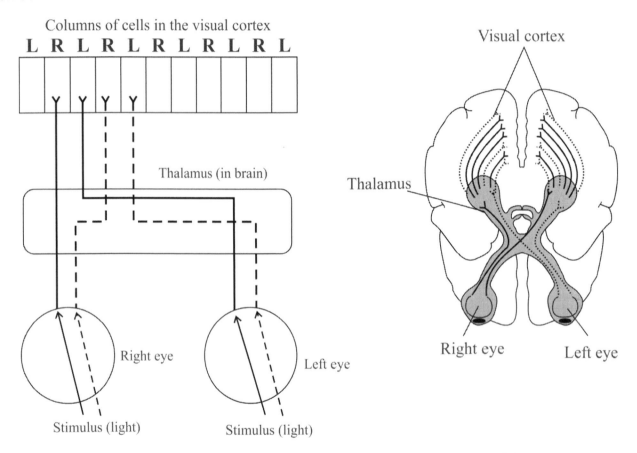

Fig. 17.2 – Development of visual cortex during binocular vision

  ✓ The columns of cells are formed before birth, but rearrangement of synapses occurs based on the visual stimulus from the environment.
  ✓ If a skill is not acquired during a critical period the acquisition of that skill in later life will be harder, even impossible.
  ✓ Visual stimulation is a very important part of development, as we receive most of our information through the eyes.

Reduced stimulation of the retina (and hence visual cortex) during the critical period for visual development in the first year or two of life can permanently damage vision.

For example, if a baby's **right eye** is injured and bandaged, light deprivation for the right eye in the baby could lead to permanently impaired vision in that eye. Synapses to cells in the visual cortex from the light-deprived eye may be lost. The columns of cells that respond to the light-deprived eye will not develop fully and the columns will be narrower, which will affect vision.

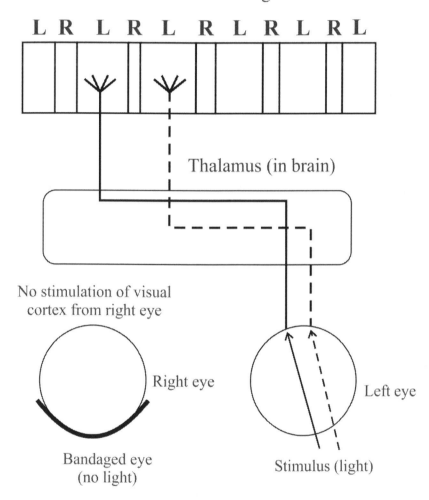

Columns of cells in the visual cortex
Wide left columns / narrow right columns

L R L R L R L R L R L

Thalamus (in brain)

No stimulation of visual cortex from right eye

Right eye

Left eye

Bandaged eye (no light)

Stimulus (light)

Fig. 17.3 – Development of visual cortex during monocular vision

Visual cortex and depth perception

Various areas of the visual cortex have been identified as being involved with the different components of vision. These areas are involved in depth perception.

Depth perception can be demonstrated in a simple experiment.

Uncap a pen and ask a fellow student to hold the cap of the pen horizontally about 15 inches away from your eye. Shut one eye and try to insert the pen into the cap, without feeling for the position of the cap. Try to use your sensory input from your eyes rather than guessing the distance of the cap. It is more difficult than it may appear. Repeat the experiment with the both eyes open. This will be a lot easier.

It is difficult to judge the distance when using a single eye (monocular vision).

Stereoscopic vision (binocular vision) allows us to obtain information about the distance of objects from us. It allows us to see objects in three dimensions. Inputs from both eyes to the cells of the visual cortex are only maintained if the neurones from each eye are stimulated at the same time and this will only happen if we use two eyes to view an object.

If one eye is covered for a long period of time, during the 'critical window', there is no input from that eye and its synapses are thus weakened and may be lost. The child may never develop stereoscopic vision. These people do not have stereoscopic vision. This can cause problems with tasks such as walking downstairs, reaching to pick things up, playing ball sports or driving a car.

Animal models to understand human brain development and function

Hubel and Wiesel's experiments with monkeys
How to process stimuli correctly must be learned. The cerebral cortex is split into column of cells. When we are born, the columns overlap and are tangled. As we learn to process stimuli, the cells organise themselves into discrete columns, which no longer overlap. There is a "critical window" for this to happen (usually before puberty, younger for visual processing). If we miss the window, our brains will become "fixed" with tangled columns and won't be able to process stimuli properly.

Hubel & Wiesel's experiments prove this.
Experiment one: Hubel & Wiesel experiments with monkeys.

Hubel & Wiesel investigated the critical window.
They used **monkeys** and **kittens** in their studies
Their work permanently blinded some animals and can be argued to be **unethical**.

Fig. 17.4 – A blind monkey

Hubel & Wiesel's Method:
1. Raise monkeys from birth in three groups for **6 months**
2. Group 1 are the control (no blindfold), Group 2 are blindfolded in both eyes, Group 3 are blindfolded in one eye (**monocular deprivation**)
3. Test the monkeys to see whether they can see using each eye
4. Test the sensitivity of retinal cells
5. Test the activity of nerves in the visual cortex in response to stimuli

The results:
- Monkeys in Group 2 (both eyes blindfolded) had impaired vision
- Monkeys in Group 3 (monocular deprivation) were blind in the deprived eye
- Retinal cells were responsive in all groups
- Cortical activity was reduced in parts of the brain that process information from the deprived eye
- Adults undergoing the same tests showed no difference between groups. All could see.

The Conclusion:
There is a critical window for visual neural development, which requires stimulus from the eye. If this window is missed the monkey is blind, because of events happening in the brain, not the eye.

Hubel and Wiesel's experiments with kittens

Experiment two: Kittens were deprived of light in one eye.

During the critical period for visual development the connections to cells in the visual cortex from the light-deprived eye have been lost and so stimuli presented through the previously deprived eye were unable to influence the majority of cells in the visual cortex. The eye that remained open during development has become the only route for visual stimuli.

Experiment three: Hubel & Wiesel experiments with kittens.

Aim: To determine if visual experience plays an active role in shaping the development of visual neurones.

Method: Kittens were placed in striped tubes (where the stripes were horizontal for some kittens and vertical for others) or were fitted with goggles that presented vertical stripes to one eye and horizontal stripes to the other.

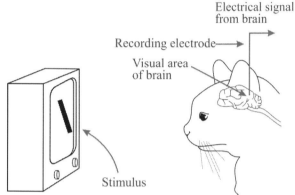

Fig. 17.5 – Visual development in kittens

Results:

When tested a few months later, after removal of the striped tubes and goggles, the kittens displayed a number of defects in their vision. They seemed to be blind to stripes with the orientation opposite to those they saw during rearing. Most of the cortical cells of the cats reared with horizontal stimuli subsequently responded strongly to horizontal stimuli and hardly at all to vertical stimuli. The opposite was true of the vertical stimuli-reared cats.

Conclusion:

These results do support the idea that visual experience affects development of cells in the visual cortex. During the critical period, cells in the visual cortex develop the ability to respond to lines presented in a particular orientation. If during that period no stimuli in one orientation are received, the 'cortical' cells are unable to respond to lines with this particular orientation later.

CHAPTER EIGHTEEN
NATURE AND NURTURE IN BRAIN DEVELOPMENT

Learning outcomes: by the end of this chapter you should be able to
Edexcel Syllabus Spec 13: *Consider the methods used to compare the contributions of nature and nurture to brain development, including evidence from the abilities of newborn babies, animal experiments, studies of individuals with damaged brain areas, twin studies and cross-cultural studies.*

It is evident from the information in the previous chapters that:
- ☆ The brain changes throughout development both structurally and functionally.
- ☆ Genes are a critical source of guidance for brain development.
- ☆ Environment-abundant shaping and fine-tuning of brain structure and function occurs with sensory-experience.
- ☆ There are critical developmental periods where nurture is essential.
- ☆ **So, in conclusion, both genetic factors (nature) and activity dependent factors (nurture) play a role in developing the brain architecture and circuitry.**

Nurture refers the environment in which development occurs.
- ○ "Nurture begins in the womb". Prenatal environments differ from child to child. Even twins will have differing prenatal environments
- ○ Experience, peer Influence, culture and child rearing practices all include nurture.

Nature refers to
- ○ The genes that we inherit from our parents
- ○ The interaction between genes (which **can be** influenced by nurture)
- ○ The expression of genes (which **can be** influenced by nurture)

Genes (nature) and / or environment (nurture)
Abilities of newborn babies
Any inborn (innate/instinctive) capacities exhibited by a newborn baby are used as evidence for the role genes in determining the hard wiring of the brain before birth. Babies are born with a range of characteristic behaviours which suggest that these are determined by genes, for example crying, smiling and grasping are all present from birth.

Experiments on depth perception using a **visual cliff** suggest that this ability too is innate.
This pioneering experiement both looks cool and is a fun exploration into the development of depth perception in creatures great and small.
In 1960, E.J. Gibson and R.D. Walk set up this awesome looking equipment that makes it look like a surface completely drops off, when actually, it's just an illusion made with shatterproof glass.
Then they put a variety of creatures on the apparatus and tried to persuade them to walk onto the glass. They used rats, cats, babies, etc and discovered that all species tested can perceive and avoid a sharp drop by the time they take up independent locomotion, be it on day one in chicks, one month old rats, or six month old babies.

Fig. 18.1 – The visual cliff

Studies of individuals with damaged brain areas

A stroke is caused due to lack of blood supply to the brain cells. Over 750,000 people suffer from a stroke in the USA annually. In fact, it is the third largest cause of death in the country. Despite this, many people recover either partially or fully from their stroke and go on to lead very healthy and happy lives. The ability of patients to recover after brain damage, for example due to a stroke, shows that neurones have the ability to respond to changes in the environment.

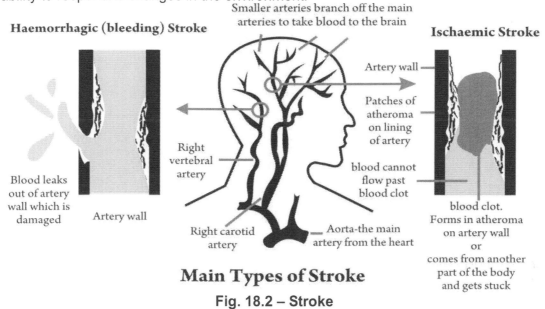

Main Types of Stroke

Fig. 18.2 – Stroke

Animal experiments

The results of experiments conducted on newborn animals have shown that development continues after birth. The failure of auditory development in birds and in children deprived of sounds also supports the idea. The role of environmental stimuli in the development of certain behaviour patterns, such as imprinting, also points to a role for nurture. The sucking behaviour of puppies is an instinctive behaviour.

An amazing and curious example of genetic and environmental influences on animal behaviour is provided the Austrian physician and naturalist Konrad Lorenz. In 1935 Lorenz reported that if he reared greylag geese from hatching, they would treat him like a parental bird. The goslings followed him about and when they were adults they courted him in preference to other greylag geese.

Fig. 18.3 – The geese followed Lorenz wherever He went

Lorenz's work provided startling evidence that there are critical periods where a definite type of stimulus is necessary for normal development. Since repeated exposure to an environmental stimulus is necessary. In nature behavioural imprinting acts as an instinct for survival in newborns. The offspring must immediately recognize its parent, because threatening events, such as the attack by a predator or by other adults could occur just after hatching. Thus, imprinting is very reliable to induce the formation of a strong social bond between offspring and parent, even if it is the wrong one. Lorenz continued these studies and in other experiments, he demonstrated that ducklings could be imprinted not only to human beings, but also to inanimate objects such as a white ball or flashing white light. He discovered also that there is a **very restricted "window" of time after hatching that will prove effective for imprinting** to take place. Lorenz was awarded the Nobel Prize for Medicine and Physiology in 1973 for this work.

Twin studies

By studying identical (monozygotic) and fraternal (dizygotic) twins and their families, scientists can estimate the relative contributions of genes and the environment to brain development. Observed differences between identical twins must be due to the effects of the environment.

If, for a characteristic, there is a greater similarity for identical twins reared together than for those reared apart, it suggests some environmental influence.

If, for a characteristic, there is a greater similarity for identical twins raised apart than for same-sex non-identical twins reared together, it suggests a strong genetic influence.

If genetic factors have a strong influence on a characteristic, then the closer the genetic relationship, the stronger will be the correlation between the individuals for that trait.

Cross-cultural studies

Studies of cultural groups who have lived in different environments with varying experiences allow the role of nature and nurture in their development to be examined. Investigations into the perception of groups from different cultural backgrounds support the idea that visual cues which most of us take for granted are not innate; they have to be learned. It suggests that visual perception is, in part, learned.

It has been argued that people in places where there are no straight roads or buildings do not have the same perception of depth as those living in a geometric world. People from 'carpentered environments' are used to interpreting acute and obtuse angles as right angles. This makes us more susceptible to optical illusions such as the Müller- Lyer and Sanders illusions, shown in Figures 18.4 and 18.5. These illusions trick your brain into thinking lines are longer or shorter than they really are.

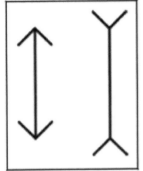

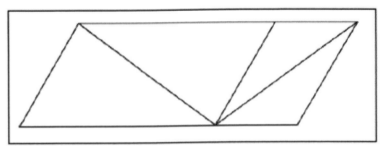

Fig. 18.4 – The vertical lines appear to be of different lengths.

Fig. 18.5 – The diagonal lines appear to be of the same length.

Are the illusions due to our innate perceptual processes (Nature), or are they affected by the environment we're brought up in (Nurture)? Cross-cultural studies have attempted to answer this question.

One study tested two completely different groups with both the illusions.

The two groups were:
- a randomly selected group of 60 black and 60 white children from a town in Illinois, USA
- black children from two different areas of Zambia – 72 from the capital town, Lusaka, and 65 from the rural environment of the Zambesi Valley.

The table shows the mean number of time the students chose the lines to be of different length, out of 10 trials.

	Group 1		Group 2	
	Illanois (Black)	Illanois (White)	Lusaka (Urban)	Zambesi Valley (Rural)
Illusion from fig. 18.4	6.1	6.1	5.8	4.6
Illusion from fig. 18.5	5.6	5.4	5.3	4.2

All of the children from Group 1 and the Lusaka children in Group 2 came from a carpentered world. These results do support the idea that people from a carpentered environment are more susceptible to illusions. The Zambesi Valley children who come from a less carpentered environment are less susceptible to the illusions. However, the difference might not be related to environment but to other factors, such as how much schooling each group of pupils had had or their ages.

People who have lived in a carpentered environment tend to interpret obtuse and acute angles as right angles because they are used to seeing them in, for example, the corners of rooms and buildings. They perceive lines with more acute angles at the top and bottom (like arrowheads) as shorter – the argument is that this is a shape you see at the nearest edge of a building to you, so you interpret the line as being shorter than it actually is. Lines that have more obtuse angles at the top and bottom (the inside-out umbrella shapes) are perceived as being longer than they actually are. This shape corresponds to what you'd see in the distant corner of a room. The mind knows that in a carpentered world rooms are usually the same height in every part of the room. Accordingly, a line with obtuse angles at its top and bottom is 'seen' as being longer than a ruler shows.

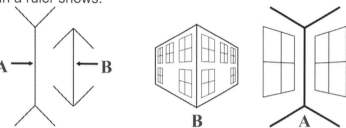

Fig. 18.6 – Geometric perception

These results do not support the hypothesis that depth perception is genetic and is positively correlated with skin pigmentation. There is little or no difference between the scores of the black and white children from Illinois. There are differences in perception between the black Illinois children, the Lusaka children and those from the Zambesi Valley. This suggests that susceptibility to the two illusions is related to differences in the environment.

CHAPTER NINETEEN
HABITUATION

Learning outcomes: by the end of this chapter you should be able to
Edexcel Syllabus Spec 14: *Describe how animals, including humans, can learn by habituation.*
Edexcel Syllabus Spec 15: *Describe how to investigate habituation to a stimulus.*

Habituation is a form of learning in which **repeated exposure to a stimulus** results in **decreased responsiveness**.
When you put your socks on you feel them at first, but after a few minutes you no longer notice them. You have become habituated to the feeling of the socks on your feet. Even though they are still providing a touch stimulus.

Habituation is not always permanent and the responsiveness will be restored if the stimulus is not available for a long period of time.

In some cases, loss of responsiveness can be due to neural fatigue or sensory adaptation rather than habituation. However, fatigue is not the cause if after a loss of one response, the same muscles can be used in another activity. Once habituated to a stimulus, an animal still senses it but has learned to ignore it.

Habituation is a very widespread and important form of learning. It enables an animal to avoid wasting time and energy responding to harmless stimuli that do not threaten its survival and reproduction.

The illustration in fig. 19.1 explains the process of habituation

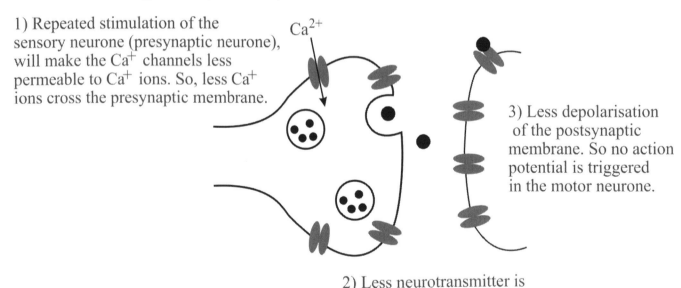

1) Repeated stimulation of the sensory neurone (presynaptic neurone), will make the Ca^+ channels less permeable to Ca^+ ions. So, less Ca^+ ions cross the presynaptic membrane.

Ca^{2+}

3) Less depolarisation of the postsynaptic membrane. So no action potential is triggered in the motor neurone.

2) Less neurotransmitter is released into the cleft

Fig. 19.1 – Habituation

Core Practical: Describe how to investigate habituation to a stimulus.

The giant sea slug breathes through a gill located in a cavity on the upper side of its body. Water is expelled through a siphon tube at one end of the cavity, if the siphon is touched, the gill is withdrawn into the cavity. This is a protective reflex action similar to removal of a hand from a hot plate. Because they live in the sea, *Aplysia* are frequently buffeted by the waves and learn not to withdraw their gill every time a wave hits them. They become habituated to waves. Habituation is a type of learning. Habituation allows animals to ignore unimportant stimuli so that limited sensory; attention and memory resources can be concentrated on more threatening or rewarding stimuli.

In this investigation students find out if habituation to a touch stimulus occurs in snails. A giant African land snail be used. The experiment is more reliable if snails are not handled much prior to the experiment, so avoiding them becoming habituated to stimuli.

1 Collect one giant African land snail and place it on a clean, firm surface. Wait for a few minutes until the snail has fully emerged from its shell and gets used to its new surroundings.

2 Dampen a cotton wool bud with water.

3 Firmly touch the snail between the eye stalks with the dampened cotton wool bud and immediately start the stopwatch. Measure the length of time between the touch and the snail being fully emerged from its shell once again, with its eye stalks fully extended.

4 Repeat the procedure in step 3 for a total of ten touches, timing how long the snail takes to re-emerge each time.

5 Record your results in a suitable table.

6 Present your results in an appropriate graph.

Fig. 19.2 explains the results that you would expect.

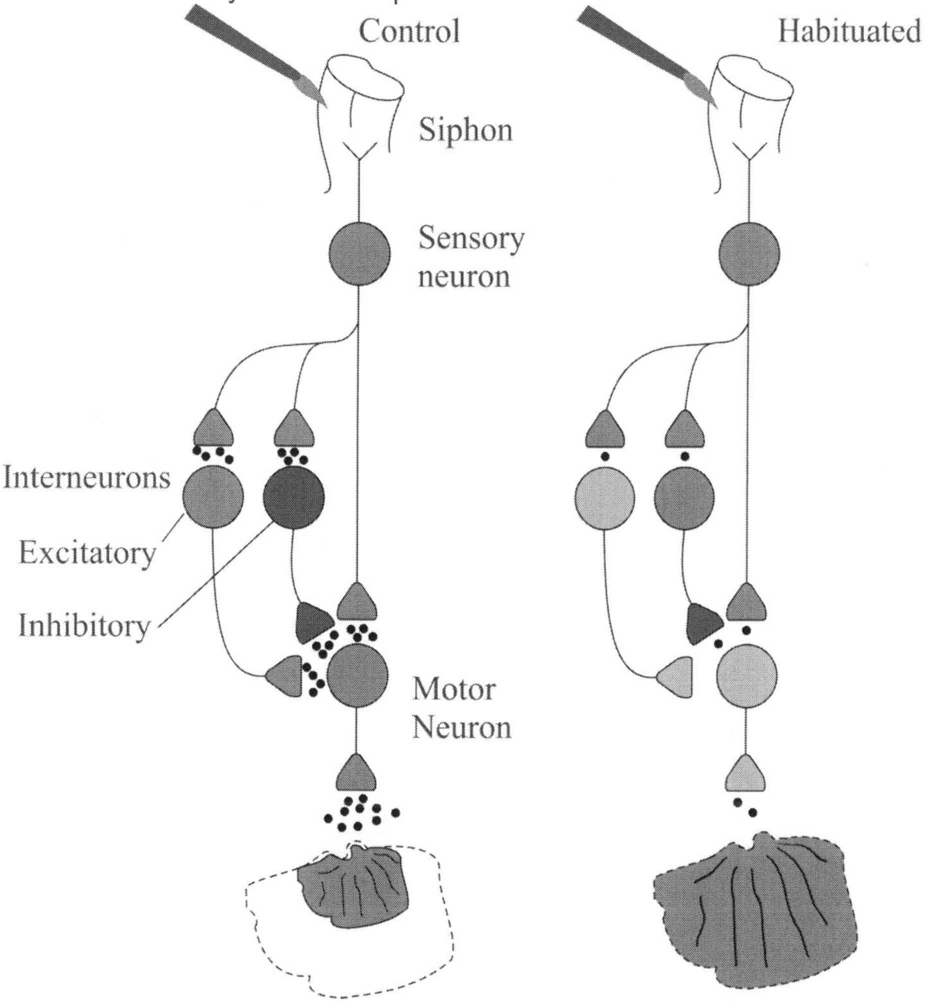

Fig. 19.2 – Habituation

Chapter Twenty
Ethics of using animals in medical research

Learning outcomes: by the end of this chapter you should be able to
Edexcel Syllabus Spec 16: Discuss the moral and ethical issues relating to the use of animals in medical research from two ethical standpoints.

Arguments For	Arguments Against
Clinical Trials Stage 1 involves animals. Without animals we would not be able to discover new drugs.	Computer simulations could be used in Clinical trials instead.
Animal testing can prevent potential loss of human life. Animals with advanced nervous systems are more likely to suffer than more primitive animals. Ability to experience pain, and self-awareness should be considered when making a choice of animals for research. Eg. The use of Daphnia is acceptable as it does not have a well developed nervous system.	Animals have rights too. Animals have no **informed consent.** Testing on animals when the potential side-effects are unknown is immoral. The species of animal may or may not affect people's perception of an animal's rights.
Utilitarian argument: Animal testing is for the greater good. Animals will also benefit as the same drugs are used to treat animals.	Animals can't tell you when and what they are suffering.
Machines like the MRI were not possible to use on animals.	Animals are often poorly cared for in labs.
Animal testing has advanced our understanding of human physiology.	Animal physiology is different to human physiology. Animal testing is, therefore, unhelpful.

Two positions on animal experiments

- **In favour of animal experiments:**
- Experimenting on animals is acceptable if (and only if):
    - suffering is minimised in all experiments
    - human benefits are gained which could not be obtained by using other methods

Against animal experiments:
- Experimenting on animals is always unacceptable because:
    - it causes suffering to animals
    - the benefits to human beings are not proven
    - any benefits to human beings that animal testing does provide could be produced in other ways

CHAPTER TWENTY ONE
PARKINSON'S DISEASES AND DEPRESSION

Learning outcomes: by the end of this chapter you should be able to
Edexcel Syllabus Spec 17: Explain how imbalances in certain, naturally occurring, brain chemicals can contribute to ill health (eg dopamine in Parkinson's disease and serotonin in depression) and to the development of new drugs.
Edexcel Syllabus Spec 18: Explain the effects of drugs on synaptic transmissions, including the use of L-Dopa in the treatment of Parkinson's disease and the action of MDMA in ecstasy.

Parkinson's disease (PD)

PD is a "multifactorial" disease that is caused by a mixture of environmental and genetic-susceptibility factors. This means that the cause of PD for most people is not found in genes or the environment alone, but in some interaction between them.

Dopamine is crucial to human movement and is the neurotransmitter that helps transmit messages that both initiate and control movement and balance. These dopamine molecules ensure that muscles work smoothly, under precise control and without unwanted movement. In **Parkinson's disease** neurons in the **frontal cortex**, brain stem and spinal cord become inactive. These neurons secrete **dopamine** neurotransmitter. This results in lack of the neurotransmitter substance dopamine in the brain. The symptoms are more intense in older people.

Symptoms
* Slowness of movement and poor balance
* Shaking of hands (tremors)
* Stiffness of skeletal muscles
* Difficulty in speaking and breathing
* **Depression**

Treatment of symptoms – Increase the levels of dopamine in the brain.
* ☆ **L-dopa**: L-dopa (levodopa) is a precursor of dopamine. Dopamine is too large to cross the blood-brain barrier in the brain. However L-dopa is small enough to cross the barrier and enter the neurones. It is then converted to dopamine and secreted. This relieves the symptoms.
* ☆ **Dopamine agonists:** these drugs bind with dopamine receptors and mimic the action of dopamine.
* ☆ **Monoamine oxidase B (MAOB) Inhibitors:** Monoamine oxidase B is an enzyme in the brain, which breaks down dopamine in the brain. **MAO B Inhibitors** are drugs which inhibit the action of MAO B. The level of dopamine remains high and the symptoms are relieved.

Depression

Depression may be caused by external factors like, relationship stress, bereavement, loss of jobs and internal factors like, low levels of serotonin and dopamine.

In **depression** neurons in the brain that secrete **serotonin** neurotransmitter stop working properly and serotonin levels fall.

Common signs and symptoms of depression
* Feelings of helplessness and hopelessness.
* Loss of interest in daily activities.
* Appetite or weight changes.
* Sleep changes.
* Irritability or restlessness.

Treatment for depression

Antidepressant medications

1. Selective serotonin reuptake inhibitors (SSRIs) are medications that increase the amount of the neurochemical serotonin at the synapses in the brain. As their name implies, the SSRIs work by selectively inhibiting (blocking) serotonin reuptake into the presynaptic neurons, as shown in fig. 21.1. This block occurs at the synapse, the place where brain cells (neurons) are connected to each other.

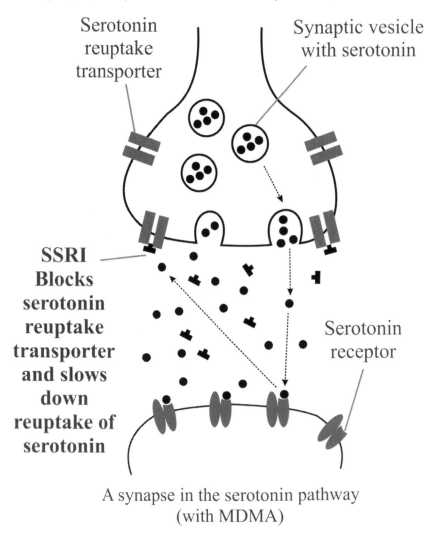

Serotonin reuptake transporter

Synaptic vesicle with serotonin

SSRI Blocks serotonin reuptake transporter and slows down reuptake of serotonin

Serotonin receptor

A synapse in the serotonin pathway
(with MDMA)

Fig. 21.1 – Selective serotonin reuptake inhibitors prevent serotonin reuptake

The SSRIs work by keeping serotonin present in high concentrations in the synapses. These drugs do this by preventing the reuptake of serotonin back into the presynaptic neurone. The reuptake of serotonin is responsible for turning off the production of new serotonin. Therefore, the serotonin message keeps on coming through. It is thought that this, in turn, helps arouse (activate) cells that have been deactivated by depression, thereby relieving the depressed person's symptoms.

2. Monoamine oxidase inhibitors (MAOIs) are the earliest developed antidepressants. MAOIs elevate the levels of neurochemicals in the brain synapses by inhibiting monoamine oxidase. Monoamine oxidase is the main enzyme that breaks down neurochemicals. When monoamine oxidase is inhibited, the neurotransmitter is not broken down and, therefore, the amount of neurotransmitter in the brain is increased.

In both cases (Parkinson's disease and depression) treatments that increase the levels of neurotransmitter might prove successful in relieving the symptoms of these diseases

Action of MDMA in ecstasy.

MDMA: MDMA is an active ingredient in ecstasy. Ecstasy is one of the most dangerous drugs threatening young people today. Called MDMA (3-4-Methylenedioxymethamphetamine) by scientists, it is a synthetic chemical that can be derived from an essential oil of the sassafras tree. MDMA is also one of the easiest illegal drugs to obtain. Its effects are similar to those of amphetamines and hallucinogens. Distributed almost anywhere, it has become very popular at social events like raves, hip hop parties, concerts, etc. frequented by both adults and youth. MDMA binds to protein pumps on the pre-synaptic membrane of nerves that secrete serotonin. The pumps would normally take serotonin up after it had been released, therefore reducing firing in post-synaptic nerves. **BUT**, when these channels are blocked, serotonin builds up in the cleft, giving greater post-synaptic activation and a sense of euphoria.

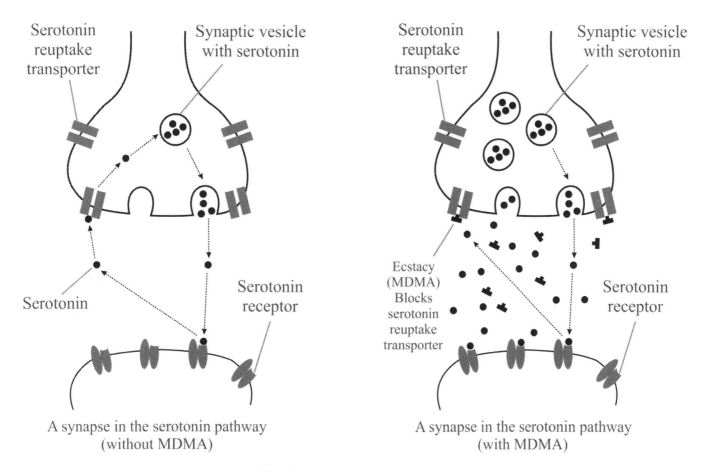

Fig. 21.2 – Action of ecstasy

Ecstasy blocks the reuptake transporter proteins in the presynaptic membrane, preventing reuptake of serotonin. This increases the concentration of serotonin in the synaptic cleft. Ecstasy molecules have a similar structure to serotonin molecules, enabling them to bind to and block the serotonin receptor sites on the transporter proteins in the presynaptic membrane. High concentrations of serotonin in the synaptic cleft will stimulate repeated impulses in the postsynaptic neurone. Neurones in the serotonin pathways stimulate the limbic system in the brain. This causes the sensations associated with 'pleasure pathways'. Pleasure pathways are intended to act as a reward system for essential survival behaviours. Stimulation of the reward pathways brought about by ecstasy will lead to repetition of the behaviour that caused the agreeable sensations. Users are more likely to be anxious, depressed and drowsy.

CHAPTER TWENTY TWO
THE HUMAN GENOME PROJECT
AND DEVELOPMENT OF NEW DRUGS

Learning outcomes: by the end of this chapter you should be able to
Edexcel Syllabus Spec 19: *Discuss how the outcomes of the Human Genome Project are being used in the development of new drugs and the social, moral and ethical issues this raises.*
Edexcel Syllabus Spec 20: *Describe how drugs can be produced using genetically modified organisms (plants and animals and micro organisms).*
Edexcel Syllabus Spec 21: *Discuss the risks and benefits associated with the use of genetically modified organisms.*

The human Genome project
A genome is all the genes found in an individual. The human genome project has sequenced the genome of about a 1000 individuals from different regions of the world. A working draft of the whole human genome was published in 2001. The sequence was immediately published on the internet, making it a freely available resource for biologists across the world.

Using the sequence biologists are
- ✓ **Gaining** a better understanding of the genome itself
- ✓ Identifying new genes
- ✓ Working out how genes are controlled
- ✓ Discovering what products genes code for.
- ✓ Understanding the role of genes in disease
- ✓ Developing new diagnostic techniques
- ✓ **Developing new drugs for treatment** of diseases.

Pharmacogenomics is the study of how an individual's genetic inheritance affects the body's response to drugs. The term comes from the words pharmacology and genomics and is thus the intersection of pharmaceuticals and genetics.

Pharmacogenomics holds the promise that drugs might one day be tailor-made for individuals and adapted to each person's own genetic makeup. Environment, diet, age, lifestyle, and state of health all can influence a person's response to medicines, but understanding an individual's genetic makeup is thought to be the key to creating personalized drugs with greater efficacy and safety.

Pharmacogenomics combines traditional pharmaceutical sciences such as biochemistry with annotated knowledge of genes, proteins, and single nucleotide polymorphisms.

Anticipated Benefits of Pharmacogenomics
- **More Powerful Medicines**: Pharmaceutical companies will be able to create drugs based on the proteins, enzymes, and RNA molecules associated with genes and diseases. This will facilitate drug discovery and allow drug makers to produce a therapy more targeted to specific diseases. This accuracy not only will maximize therapeutic effects but also decrease damage to nearby healthy cells.
- **Better, Safer Drugs the First Time:** Instead of the standard trial-and-error method of matching patients with the right drugs, doctors will be able to analyze a patient's genetic profile and prescribe the best available drug therapy from the beginning. Not only will this take the guesswork out of finding the right drug, it will speed recovery time and increase safety as the likelihood of adverse reactions is eliminated.

- **More Accurate Methods of Determining Appropriate Drug Dosages:** Current methods of basing dosages on weight and age will be replaced with dosages based on a person's genetics -- how well the body processes the medicine and the time it takes to metabolize it. This will maximize the therapy's value and decrease the likelihood of overdose.
- **Advanced Screening for Disease** will allow a person to make adequate lifestyle and environmental changes at an early age so as to avoid or lessen the severity of a genetic disease.
- **Better Vaccines:** Vaccines made of genetic material, either DNA or RNA, promise all the benefits of existing vaccines without all the risks. They will activate the immune system but will be unable to cause infections. They will be inexpensive, stable, easy to store, and capable of being engineered to carry several strains of a pathogen at once.

One of the main project goals of the Human Genome Project is to address the ethical, legal and social issues that may arise from the project. However, many issues are still to be addressed adequately:

- Testing for genetic predisposition has many implications, would it be acceptable, for example, for insurers to have this information about people who are applying for health insurance.
- Who should decide about the use of genetic predisposition tests, and on whom they should be used?
- Making and keeping records of individual genotypes raises acute problems of confidentiality.
- Many medical treatments made possible through the development of genetic technologies will initially be very expensive. Their restricted availability will add considerably to the problems faced by the health services in deciding who is eligible for such treatments.

Drug production by using genetically modified organisms

Gene technology enables scientists to manipulate DNA in many ways.

- ❖ Individual genes can now be identified in the DNA of an organism. An individual gene can be isolated, removed and cloned.
- ❖ The DNA of an organism can be combined with the DNA of another.
- ❖ Genes (DNA) can also be made from the RNA of an organism.

Some of the tools used in gene technology are tabulated below:

TECHNIQUE	PURPOSE
Restriction Enzymes	To cut DNA at specific points, making small fragments
DNA Ligase	To join DNA fragments together
Vectors	To carry DNA into cells and ensure replication
Plasmids	Common kind of vector
Genetic Markers	To identify cells that have been transformed
PCR	To amplify very small samples of DNA
cDNA	To make a DNA copy of Mrna
DNA probes	To identify and label a piece of DNA containing a certain sequence
Gene Synthesis	To make a gene from scratch
Electrophoresis	To separate fragments of DNA
DNA Sequencing	To read the base sequence of a length of DNA

All these procedures are possible only with the use of different enzymes. Three important enzymes and their roles in gene technology are shown.

a) **Reverse transcriptase** ; This enzyme is used to make DNA (genes) from RNA. First, the mRNA is extracted from the cell. The mRNA is a mirror image copy (complementary) of the desired gene.

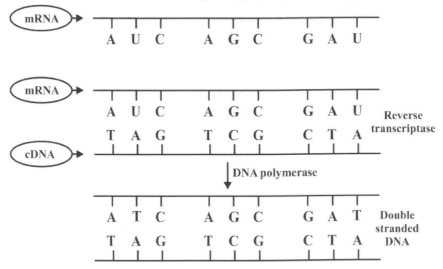

Fig. 22.1 – Syntheising a gene from mRNA

❖ The enzyme reverse transcriptase is then used to form a single strand of **complementary DNA (cDNA)**, by using mRNA as a template.
❖ The single stranded cDNA can then act as a template to form a double stranded DNA helix of the gene. The enzyme DNA polymerase is used to make a double stranded DNA from a single stranded of cDNA.

b) Restriction Endonuclease: These are enzymes that cut DNA into small fragments. This allows individual genes to be isolated. This enables a gene from one organism to be inserted into the DNA of another organism.
❖ The restriction endonuclease makes a staggered cut in the DNA molecule between specific base sequences. There are hundreds of different restriction endonucleases, each cuts DNA at a specific base sequence. Example of the restriction endonucleases ECORI will always cut DNA between the bases as shown.

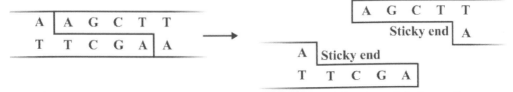

Fig. 22.2 – Restriction endonucleases cut DNA at restriction sites
The specific nucleotide sequence between which the endonuclease cuts the DNA is called a **recognition site or restriction site**. The cut ends are complementary to each other and will have a natural affinity for each other. These are referred to as 'Sticky ends'.

c) DNA ligase: Once the donor DNA and recipient DNA have been cut by the same restriction endonuclease, the sticky ends are joint up using the enzyme DNA ligase. This helps to form **Recombinant DNA.**

❖ **Plasmids are circular loops of extra-chromosomal DNA in bacterial cells. They sometimes contain genes for antibiotic resistance. This can be used as 'markers'. The small size of plasmids and their ability to replicate within bacterial cells, makes them good genetic tools to transfer genes into bacterial cells.**

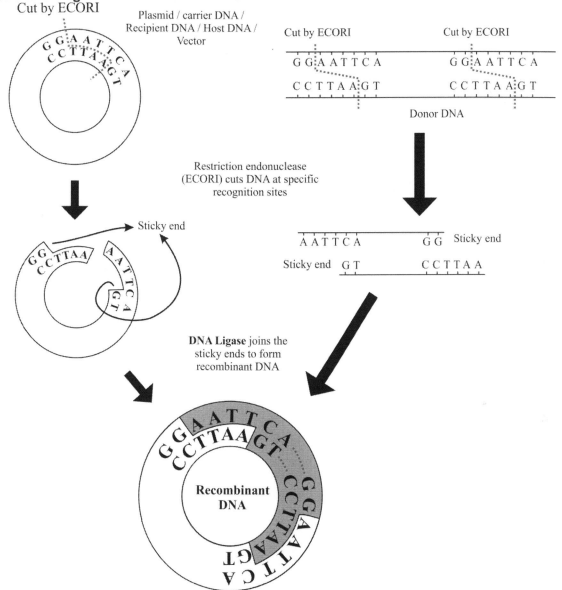

Fig. 22.3 – formation of recombinant DNA

Vectors containing the genes we want must be incorporated into living cells so that they can be replicated or expressed. The cells receiving the vector are called <u>host cells</u>, and once they have successfully incorporated the vector they are said to be <u>transformed</u>. Vectors are large molecules which do not readily cross cell membranes, so the membranes must be made permeable in some way.

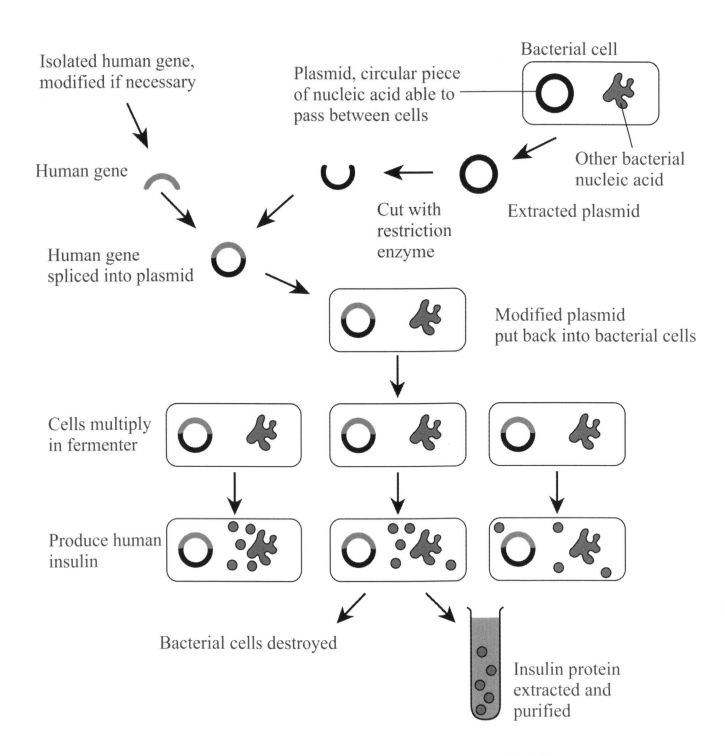

Fig. 22.4 – Producing Insulin by using Genetically Modified Bacteria

Introducing pharmaceutical Genes into Crop Plants (Using Agrobacterium tumefaciens)

The bacterium *Agrobacterium tumefaciens* enters into plant cells through wounds. The plasmid of this bacterium has a tumor-introducing gene. This gene is inserted into the DNA of one of the plant chromosomes. The gene then causes tumor formation in plants. These tumors are called "Crown Gall". The plasmid of *Agrobacterium tumefaciens* is called the T_1 plasmid.

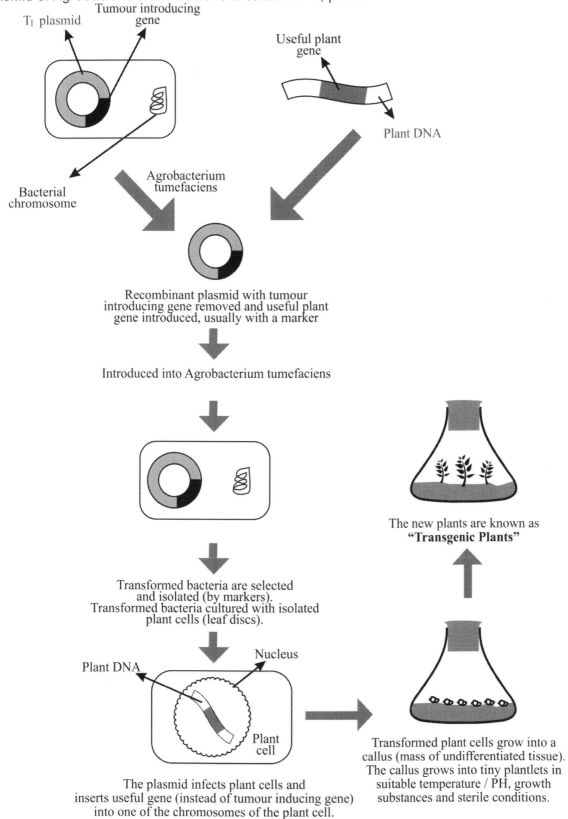

Fig. 22.4 – Producing transgenic plants

Genetically modified crops such as maize or soya have the potential to mass produce medicines and other chemicals cheaply and efficiently. Plants in crop trials have been genetically engineered to manufacture proteins for healing wounds and for treating conditions such as cirrhosis of the liver, anaemia and cystic fibrosis (CF). Other trials are exploring the possibilities of using genetically engineered plants to produce antibodies to fight cancer, and vaccines against rabies, cholera and foot-and-mouth disease.

Introducing genes into animals

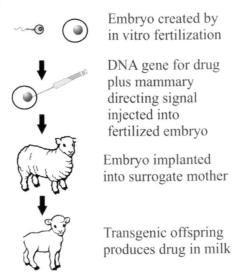

Embryo created by in vitro fertilization

DNA gene for drug plus mammary directing signal injected into fertilized embryo

Embryo implanted into surrogate mother

Transgenic offspring produces drug in milk

Fig. 22.5 – Production of transgenic animals for pharmaceutical uses

Products such as insulin, growth hormone, and blood anti-clotting factors may soon be or have already been obtained from the milk of transgenic cows, sheep, or goats. Research is also underway to manufacture milk through transgenesis for treatment of debilitating diseases such as phenylketonuria (PKU), hereditary emphysema, and cystic fibrosis.

In 1997, the first transgenic cow, Rosie, produced human protein-enriched milk at 2.4 grams per litre. This transgenic milk is a more nutritionally balanced product than natural bovine milk and could be given to babies or the elderly with special nutritional or digestive needs. Rosie's milk contains the human gene alpha-lactalbumin.

Risks and benefits associated with the use of genetically modified organisms

Potential benefits	Potential risks
Could help to feed the developing world • Approximately one third of crops worldwide are lost to disease and pests, and so GM crops resistant to disease and pests would increase yield. GM crops could increase yield in hostile conditions such as drought, and extend the amount of land available for agriculture.	**Will not be able to feed the world** • Many people argue that worldwide hunger issues are not based on an inability to produce enough food. There is enough food to go around if it was distributed equally. It's more a problem of politics, poverty and trading, which will not be solved using GM crops.
GM crops are more cost-effective • GM crops that produce chemicals with altered composition (e.g. starch and oils) could reduce costs in manufacturing certain products.	**Could damage organic farmers** • If GM crops are able to crossbreed with organic crops, then the organic crops are no longer considered to be organic. This would damage the livelihood of organic farmers.

Could benefit human health

GM crops could be modified to contain modified nutritional contents that would have health benefits. The GM Golden Rice – this has a high vitamin A content, and reduces the number of people suffering Night blindness.

Farmers producing food in developing countries sometimes use pesticides that are banned in developed countries. This can damage their health. If GM crops didn't need pesticides, this damage to health would not happen.

GM crops (pharma crops) could allow cheap production of drugs, including ones which are currently very expensive, and so unlikely to be widely available particularly in developing countries. There's the potential to produce GM crops containing vaccines for certain conditions. This may allow people access to vaccines without specialised storage (such as refrigeration).

May have unpredictable health risks

We cannot know the consequences of transferring genes across the species barrier. Proteins produced may be dangerous (toxic or initiate allergic responses), either directly, or indirectly through metabolic processes. Pharma crops may breed with conventional crops, meaning that there are potentially harmful chemicals entering the food chain.

Could reduce pesticide and herbicide use

Reduced chemical use would be beneficial to health and reduce the cost of crop production. Herbicide-tolerant crops would allow one type of herbicide to be used all year round, reducing the need for a complex mixture of herbicides to be applied at different times of year. Also, crops able to produce their own insecticides (e.g. Bt toxin) reduce the need for pesticides.

Could increase herbicide and pesticide use

The genes used for increased herbicide resistance in GM crops may spread to other plants, including weeds. This may allow for the development of resistant 'superweeds'. Therefore, there would be increased herbicide use to kill the 'superweeds'. GM crops able to produce insecticides could allow insects to develop resistance. If the insects are exposed to a constant high dose of an insecticide (a selection pressure), mutants resistant to the insecticide would be selected for. This would allow 'superbugs' to develop, requiring an increasing use of chemicals on crops.

Could help preserve natural habitats

Reduced use of pesticides and herbicides could mean that there's less impact on surrounding habitats. Currently, natural habitats are being cleared to allow more space for crops with a low yield. If higher yield crops were introduced, there would be no need to clear more habitats for crops.

Could reduce biodiversity

It has been argued that the use of GM crops reduces the chance of other plants or insects existing within the crop, and hence reduces biodiversity. However, it is argued that this is essentially the same as farming a specific crop. Superweeds resulting from transfer of genes from GM crops to weed species could result in weeds outcompeting native wild species.

	Mainly benefits big biotech companies Biotech companies have the opportunity to patent genes, techniques and crops they develop. This means that any profit generated by this technology is owned solely by one company. These companies are private companies, who are interested in the technology to make profit, not necessarily for the greater good of humanity.
	Raises ethical conflicts over the control of food production As big biotech companies hold patents for these technologies, they have sole control. This may mean that they have too much influence over food production in developing countries. Many crops generated are infertile, and so farmers are unable to keep seeds from a crop in order to plant the following year. Therefore, if they want to grow the same GM crop the following year they need to buy seeds again, which many farmers in developing countries cannot afford.

6239391R00050

Printed in Great Britain
by Amazon.co.uk, Ltd.,
Marston Gate.